Automotive Oscilloscopes Setup and Use

Welcome to Automotive Oscilloscopes Setup and Use

This book has been written as a companion to 🔍 *Automotive Oscilloscopes Waveform Analysis*.

As electrics and electronics continue to develop in modern vehicle design, technicians need to leverage the power of advanced diagnostic tools to remain current and effective in the field of automotive repairs.

The power of an oscilloscope to provide insights into the hidden world of electrical systems and components is unparalleled; it's like having X-ray vision. It is the oscilloscope (or scope) that can truly test the operation and health of a system component.

An effective diagnostic routine should always begin with a logical assessment of symptoms and then use reasoning to reduce the possible number of options before following a systematic approach to finding and fixing the root cause.

An oscilloscope can help you accurately assess these symptoms.

An important thing to remember about oscilloscopes is that they should be easy to set up and use; otherwise, they'll be passed over for a more familiar tool within your comfort zone.

Remember that nothing ever happens within your comfort zone.

There is a great deal of misconception about how difficult a scope can be to set up, and once you are used to your own equipment, if it is laid out and ready to use, it will soon become your diagnostic tool of choice.

This book has been written to help you get the most from your oscilloscope and has been designed to give both straightforward and uncomplicated methods that can be used effectively for automotive diagnosis and more advanced methods to help reinforce and expand on the concepts explained.

The chapters will introduce you to health and safety, electrical principles and the setup and use of oscilloscopes, including key terms, points of interest and diagnostic tips to support the information provided within the text.

Chapters

How to use this book..........................**Page 3**

Chapter 1 Electrical Essentials................**Page 7**

Chapter 2 Introduction to Oscilloscopes....**Page 27**

Chapter 3 Setup and Use......................**Page 49**

Chapter 4 Oscilloscope Technology and Features……………………………………...**Page 93**

Chapter 5 Mastering Signal Analysis…......**Page 114**

Chapter 6 Advanced Techniques………....**Page 170**

This book offers:

Information to help Automotive Technicians systematically diagnose electrical and electronic vehicle faults using an oscilloscope.

Ideal support for learners and tutors undertaking automotive qualifications.

A large number of illustrations to support knowledge and understanding, with an analysis of automotive waveforms.

Text © Graham Stoakes 2025

Original illustrations © Graham Stoakes 2025

The rights of Graham Stoakes to be identified as author of this work have been asserted by them in accordance with the Copyright, Designs and Patents Act 1988.

Automotive Oscilloscopes
Setup and Use

Copyright notice ©

All rights reserved. No part of this publication may be reproduced in any form or by any means (including photocopying or storing it in any medium by electronic means and whether or not transiently or incidentally to some other use of this publication) without the written permission of the copyright owner, except in accordance with the provisions of the Copyright, Designs and Patents Act 1988 or under the terms of a licence issued by the Copyright Licensing Agency, Saffron House, 6 - 10 Kirby Street, London EC1N 8TS (www.cla.co.uk). Applications for the copyright owners' written permission should be addressed to the author.

Acknowledgements

Graham Stoakes would like to thank Anita and Holly Stoakes for their support during this project.

Thank you to alerrandre for the cover design.

Thank you to Nathan Ross and Jamie Cushenan of No More Copyright for help with images.
https://nomorecopyright.com/

The author and publisher would also like to thank the following individuals and organisations for permission to reproduce photographs:

Cover image: No more copyright

ProMoto Europe

Author

Graham Stoakes AAE MIMI QTLS is a trainer/lecturer and author of college textbooks in automotive engineering for light vehicles and motorcycles.

With his background as a qualified Master Technician, senior automotive manager, and specialist diagnostic trainer, he brings over 40 years of technical industry and training experience to this title.

www.grahamstoakes.com

Cover design - fiverr.com/alerrandre

Published by Graham Stoakes

First published 2025

First edition

ISBN 978-0-9929492-9-7

Some of the waveforms, analysis and tests contained in this book are based around the features included with the industry leaders PicoScope, one of the most popular automotive oscilloscopes available.

More information and some really useful additional resources can be found on the Pico Technology website www.picoauto.com.

Introduction

How to use this book

Are you new to oscilloscopes, or do you want to enhance your current knowledge?

If so, then you are ready to use an oscilloscope for automotive electronic diagnosis.

The oscilloscope allows you to view the normally hidden worlds of current, voltage, pressure, and much more. Once you become familiar with the function and operation of automotive electrical circuits and components, you'll be able to use an oscilloscope to effectively analyse waveforms and diagnose faults.

This book is fully illustrated to provide detailed information about the setup, use, and procedures utilised when conducting diagnosis with an oscilloscope. Although it is impractical to address every feature, setup, and test available on oscilloscopes, this book aims to offer a selection and overview of various styles. This approach is intended to help you build the confidence necessary to explore more advanced routines.

Due to the wide range of circuits, components, and operating faults possible on a vehicle, this publication attempts to remain generic, allowing you the flexibility to adapt the content to your diagnostic needs without tying you down to specific equipment, settings or figures. It concentrates more on the setup of the procedures and functions available.

Throughout this book, you will find features that aim to support and enhance your understanding and use, such as:

The information in these boxes highlights safety features to consider when working on vehicles and electrical circuits, especially high-voltage systems. These features aim to minimise the risk of injury or damage to vehicles or equipment. Even if specific safety advice is not provided, always evaluate potential risks before starting any activity or diagnostic routine.

The guidance in these boxes is intended to support the information about the construction and operation of vehicle systems and diagnosis. It provides material that enhances understanding and strengthens knowledge of system components and testing methods.

This feature explains the key terms related to oscilloscopes, vehicle operation, components, and diagnostic testing. Understanding and correctly using technical vocabulary is the foundation for effective repairs. Words highlighted in **bold** within the text are defined here.

Introduction

 These tips offer useful diagnostic advice for specific systems and components. Although not all of them will be relevant to your current task or vehicle, they may inspire ideas that you can modify and incorporate into your diagnostic routines. Always take care when implementing any diagnostic process to avoid the possibility of damage or injury to yourself, others, vehicles, or equipment.

 The guidance in these boxes uses analogies to compare complex operational systems or component designs to simple concepts. The purpose of these explanations is to clarify and improve understanding, even though they are not scientifically accurate. However, they are only a tool to aid comprehension, not a replacement for correct information.

Preparing for assessment

The information in this book can help you with theory or practical assessments that measure your skills or competence in vehicle repairs or a recognised qualification. You may be able to use some of the evidence you produce for more than one qualification. You should make the best use of all your evidence to maximise the opportunities for cross-referencing between units or qualifications.

You should choose the type of evidence that suits the type of assessment you are undertaking (either theory or practical).

The types of evidence you could use are listed below:

- Direct observation by a qualified assessor
- Witness testimony
- Computer-based
- Audio recording
- Video recording
- Photographic recording
- Professional discussion
- Oral questioning
- Personal statement
- Competence/Skills tests
- Written tests
- Multiple-choice tests
- Assignments/Projects

Before taking a written or multiple-choice test, review the key terms related to the subject. Read all questions and answers thoroughly to understand what is being asked, as multiple-choice tests often have similar options that can be confusing.

For practical assessments, make sure you have had ample practice and feel confident in your ability to pass. Having a plan of action and a work method can be helpful.

Ensure you have the correct technical information, such as vehicle data, and the necessary tools and equipment. Check your work regularly to ensure accuracy and prevent issues from developing.

Always prioritise safety when performing any practical task.

Introduction

Information sources

The complex nature of light vehicle electric and electronic systems requires a good source of technical information and data. In order to conduct diagnostic, maintenance and repair procedures, you need to gather as much information as possible before you start.

Sources of information may include:

Table 0.1 Possible information sources

Verbal information from the driver	Vehicle identification numbers
Service and repair history	Warranty information
Vehicle handbook	Technical data manuals
Workshop manuals/Wiring diagrams	Safety recall sheets
Manufacturer specific information	Information bulletins
Technical helplines	Advice from other technicians/colleagues
Internet	Parts suppliers/catalogues
Jobcards	Diagnostic trouble codes
Oscilloscope waveforms	On-vehicle warning labels/stickers
On-vehicle displays	Reference/Textbooks

Always compare the results of any inspection, testing or diagnosis to suitable sources of data. Remember that no matter which information or data source you use, it is important to evaluate how useful and reliable it will be to your safety, diagnostic, maintenance and repair routine.

Electronic and electrical safety procedures

Working with any electrical system has its hazards, and you must take safety seriously. When working with light vehicle electrical and electronic systems, the main hazard is the risk of electric shock. Although most systems operate with low voltages of around 12V, an accidental electrical discharge caused by incorrect circuit connection can be enough to cause severe burns. Where possible, isolate electrical systems before repairing or replacing components.

If working on hybrid or fully electric vehicles, take care not to disturb the high voltage system. The high voltage system can normally be identified by its reinforced insulation and shielding, often coloured bright orange. These systems carry voltages that can cause severe injury or death.

Always use the correct tools and equipment. Damage to components, tools or personal injury could occur if the wrong tool is used or misused. Check tools and equipment before each use.

If you are using electrical measuring equipment, check that it is correctly rated, accurate, and calibrated before you take any readings.

If you need to replace any electrical or electronic components, always check that the quality meets the original equipment manufacturer (OEM) specifications. (If the vehicle is under warranty, inferior parts or deliberate modification might make the warranty invalid. Also, if parts of an inferior quality are fitted, it might affect vehicle performance and safety). You should only carry out the replacement of electrical components if the parts comply with the legal requirements for road use and environmental protection.

Introduction

 Although oscilloscopes can sometimes be used for testing the high voltage systems of hybrid and electric vehicles, you <u>should not</u> attempt diagnosis and repair of these vehicle types unless you have had specific training and are using the correct high voltage Personal Protective Equipment (PPE).

Personal Protective Equipment (PPE)

To reduce the possibility of personal injury, always use the appropriate personal protection equipment (PPE):

When selecting PPE, make sure that the equipment:

- Is the right PPE for the job – ask for advice if you are not sure.
- Fits correctly – it needs to be adjustable, so it fits you properly.
- Is properly looked after.
- Prevents or controls the risk for the job you are doing.

- Does not create a new risk, e.g. Overheating.
- Is comfortable enough to wear for the length of time you need it.
- Does not impair your sight, communication or movement.
- Is compatible with other PPE worn.
- Does not interfere with the job you are doing.

Vehicle Protective Equipment (VPE)

To reduce the possibility of damage to the car, always use the appropriate vehicle protection equipment (VPE):

  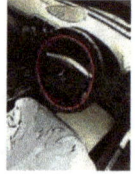 

Wing covers   Seat covers   Steering wheel covers   Floor mats

Electrical Essentials

Chapter 1 Electrical Essentials

In order to make effective use of an automotive oscilloscope, it is important to have a foundational knowledge of electrical essentials. This chapter will help you develop an understanding of fundamental electrical principles used in automotive engineering. It also introduces the basic operating theories of electricity and electrical systems that will aid you when undertaking maintenance and repairs. Remember to work safely at all times and observe the relevant health and safety regulations, while developing diagnostic routines that are systematic and effective.

Contents

What is Electricity	**8**
Electrical Units and Terminology	**11**
Circuit Properties and Voltage Types	**16**
Electromagnetism	**18**
Chemical Electricity	**18**
Alternating Current AC and Direct Current DC	**19**
High and Low Voltage	**22**
Ohms and Watts (Power) Law	**23**
Pulse Width Modulation and Duty Cycle	**25**
Pull-up and Pull-down Circuits	**26**

The automotive industry is a high-risk environment, especially when dealing with electrical systems. The hazards of electricity are well-known but can be easily ignored due to its invisible nature. This can lead to complacency if the fundamentals of electricity are not well understood. Even with this understanding, caution is necessary. Assume that any safety systems designed for protection have failed and take precautions to minimise the risk of injury or death. Always evaluate the risks associated with any activity and implement measures to eliminate or reduce the hazards involved in any task, diagnosis, or repair. Additional risks associated with working on, or around electrical systems may include:

- Electrocution
- Strong magnetic fields
- Falling from heights
- Short circuits
- Electrical discharge/arcing
- Fire and explosion
- Chemicals

Electrical Essentials

What is Electricity

The discovery of electricity

Approximately 2,500 years ago, a Greek scientist named Thales discovered that rubbing amber (fossilised tree sap) with a cloth attracted small dust and fluff particles. This was his discovery of static electricity. While Thales did not fully understand the phenomenon, he did document his findings.

Around 1550, William Gilbert, who was Queen Elizabeth I's doctor, discovered that rubbing a silk cloth on a glass rod could attract even heavier objects, like feathers. He called this phenomenon 'electricus', taking the name from the Greek word for amber, 'elektron', leading to the word electricity.

While static electricity is interesting, it's hard to convert into a usable energy source because electricity needs to move to be useful. In the late 18th century, two Italian scientists, Luigi Galvani, and Alessandro Volta, were competing with each other and ended up creating the first moving electricity, known as electric **current**. This electric current was produced through a chemical reaction and eventually led to the invention of the battery.

 Understanding electricity can be challenging because it exists within tiny atoms. You can picture an atom as a miniature solar system, where the **nucleus** is like the sun and the electrons orbit it like planets. The nucleus contains positively charged protons and neutral neutrons, while the electrons that orbit it have a negative charge. When electrons move from one atom to another, they create electric current.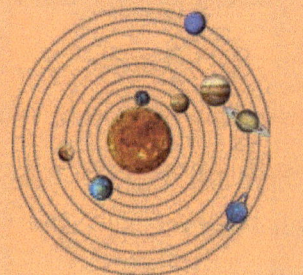

Atoms and molecules

Every substance is composed of **molecules**, which are made up of **atoms**. For instance, water is a molecule denoted as H_2O, comprising two hydrogen (H) atoms and one oxygen (O) atom.

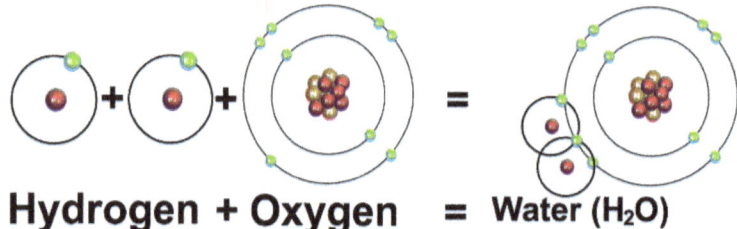

Hydrogen + Oxygen = Water (H_2O)

Figure 1.1 Hydrogen and oxygen making up a water molecule

Current - the flow of electric charge though a conductor.

Nucleus - the central part of an atom.

Atom - the smallest unit of matter that has the characteristic properties of a chemical element.

Molecule - a group of two or more atoms that are held together by chemical bonds.

Electrical Essentials

The number of **protons** and **electrons** varies among atoms, as depicted in the periodic table. This chart organises elements by atomic number, which mainly corresponds to the number of protons in their nucleus.

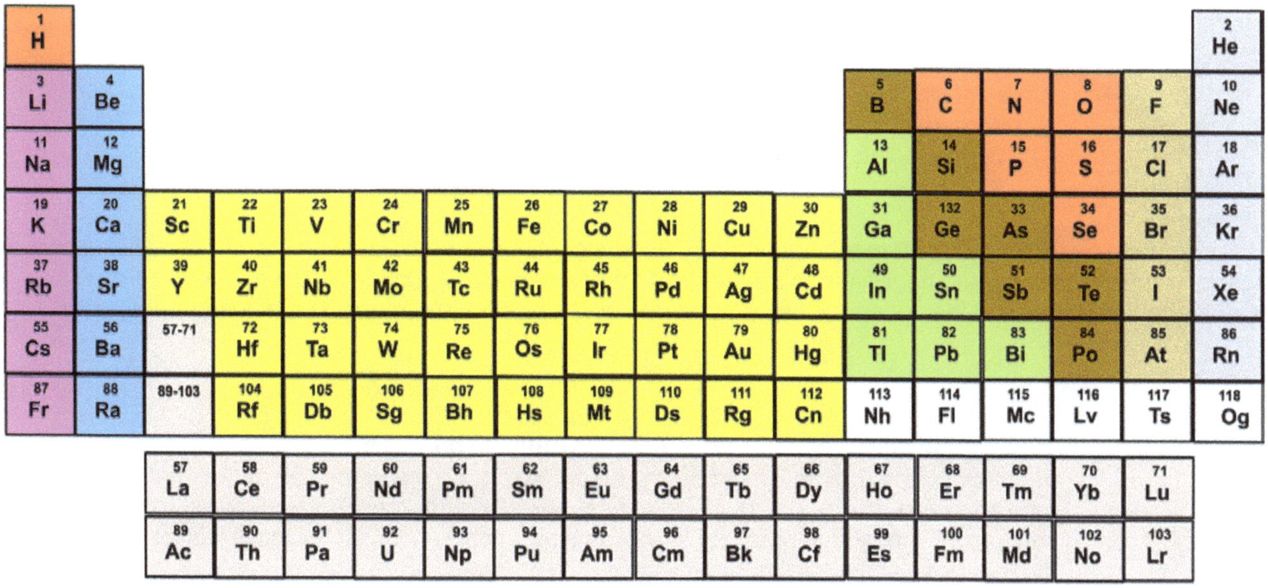

Figure 1.2 The periodic table of elements

Movement of electrons

To generate an electric current, electrons need to move from one atom to another. Electron movement requires an external force or pressure, which can be created by magnetic fields or a chemical reaction.

To generate an electric current, electrons must move from one atom to another. This movement needs an external force or pressure, which can be created by magnetic fields or chemical reactions.

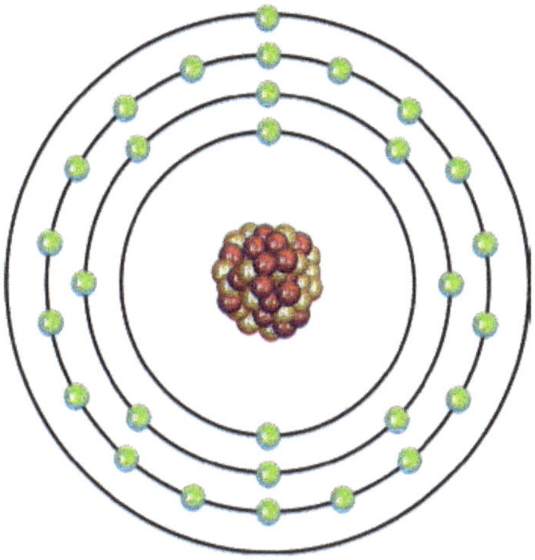

Electrons orbit the nucleus of an atom, much like planets orbit the sun due to gravity. In atoms with simple structures, the attraction between the nucleus and electrons is very strong, making it difficult for electrons to move. Elements where electrons do not move easily are known as **insulators**.

However, in other atoms, the attraction force between the nucleus and electrons is weaker. For example, a copper atom, which has 29 electrons and 29 protons. The electrons orbit in increasingly larger circles. The outermost electrons, known as 'free electrons', have a weaker bond to the nucleus than those in simpler atoms. If external pressure is applied, these free electrons can be made to move from one atom to another, creating an electric current. When electrons move easily, the element is known as a **conductor**.

Figure 1.3 A copper atom

Electrical Essentials

In vehicles, conductors are used where we want electricity to flow easily, such as in wiring. Insulators are used to restrict the movement of electricity, such as the coating on the outside of a high-voltage cable.

 Certain elements, such as silicon and germanium, can be engineered into components that function as either conductors or insulators. They can even be switched between these two states, acting as controls in electronic systems. These versatile elements are known as **semiconductors**.

For electrons to move from one atom to another, they need a continuous path, known as a **circuit**. This allows an electron to be replaced by another one from behind as it moves. Without a complete circuit, electrons can't flow because the last electron in the conductor has nowhere to go. If the circuit is interrupted, it loses **continuity**.

 Imagine a relay race where runners pass a baton to one another in a continuous loop around the track. If at any point a runner drops the baton or doesn't show up at the next handoff, the race comes to an abrupt stop. In the same way, electrical continuity ensures that electrons can 'pass the baton' seamlessly in a complete, unbroken circuit. Without this unbroken path, the 'race' of electric current halts entirely.

Proton - a subatomic particle that has a positive electric charge and is found in the nucleus of every atom.

Electron - a subatomic particle that has a negative electric charge and is one of the main components of matter.

Insulator - an electrical component that restricts or prevents the flow of electric current.

Conductor - an electrical component which allows the flow of electric current.

Semiconductor - a material that can have the properties of both a conductor and an insulator when used in an electric circuit.

Circuit - a continuous, unbroken loop that allows the steady flow of electricity.

Continuity - refers to a complete, unbroken conductor that enables the uninterrupted flow of electricity.

Electrical Essentials

Electrical Units and Terminology

To better understand how electricity works in a circuit, we use specific units of measurement. Correct use of electrical terminology is crucial to avoid misinterpretation, which could lead to incorrect testing and diagnosis.

Table 1.1 outlines the primary units associated with electricity and the function of electrical systems.

Table 1.1 Units of measurement

Unit	Description
Volts	Voltage, named after Alessandro Volta, is the electrical pressure or potential force in any part of an electrical circuit. There are two main types of voltage in electrical circuits: • **Electromotive Force (EMF)**: This is the potential pressure when all electrical devices are turned off and no current is flowing, often considered as the **open circuit voltage (OCV)**. • **Potential Difference (Pd)**: This is the voltage drop caused by electricity flow when the circuit is active or switched on. When the circuit is active, this is known as the **closed-circuit voltage (CCV)**. Voltage is often represented in technical information or documentation as: V - Volts named after Alessandro Volta. E - EMF to describe electromotive force.
Amps	Amps, named after André-Marie Ampère, are the units used to quantify the amount of electricity in any part of an electrical circuit. It's measured when electricity is allowed to flow in a circuit, a phenomenon known as current. There are two main types of electrical current: • Direct Current (DC): This is electricity that flows in one direction only. • Alternating Current (AC): This is electricity that **oscillates** back and forth in a circuit. Regardless of where you measure it in the circuit (at the beginning, middle, or end), the amperage remains the same. Current (amps) is often represented in technical information or documentation as: I - which originates from the French phrase intensité du courant, (current intensity). A - amps, named after André-Marie Ampère.

Electrical Essentials

Table 1.1 Units of measurement

Unit	Description
Ohms 	Ohms, named after Georg Ohm, are the units used to measure **resistance** to electrical flow. Resistance directly impacts the functioning of any electrical circuit as it slows down the flow of electricity. As resistance increases in a circuit, both current and voltage decrease. This can limit the operation of electrical components. While resistance can be used to control electrical components in some circuits, high resistance is generally undesirable. Resistance in electrical circuits is often closely associated with heat. Therefore, heat can be seen as an indication of resistance, and vice versa. Resistance (ohms) is often represented in technical information or documentation as: Ω - the Greek letter Omega (meaning 'great'), which sounds similar to ohms and ensures that the letter 'O' is not confused with a zero.
Watts 	Watts, named after James Watt, are the units used to measure electrical power produced or consumed. **Power** is essentially the speed at which work is done. In the context of electrical components, a higher wattage indicates a more powerful component that uses more electrical energy. Power is often represented in technical information or documentation as: W - Watts, named after James Watt. P - to represent the word power. The word horsepower is often attributed to James Watt but considered to be an imperial measurement. The Si unit of power is the Watt. 1 horsepower (hp) is equivalent to approximately 746 Watts. 1.34 horsepower (hp) equals 1 kilowatt (kW).
Hertz 	Hertz, named after Heinrich Hertz, are the units used to measure **frequency**. Frequency relates to how often something happens in one second of time. In automotive terms, it can be used to describe things like the frequency of electrical signals, vibrations, or oscillations. For example, if a sensor operates at 1,000 Hz, it means it cycles or measures 1,000 times per second. Frequency is often represented in technical information, documentation or on oscilloscope displays as: Hz - Hertz, named after Heinrich Hertz.

Electrical Essentials

Table 1.1 Units of measurement

Unit	Description
Coulombs 	Coulomb, named after Charles-Augustin de Coulomb, is a unit of electric **charge**. It tells us how much electric charge is present or transferred in a system. One coulomb is the amount of charge carried by approximately 6,241,509,000,000,000,000 electrons. Charge is often represented in technical information or documentation as: C - Coulombs, named after Charles-Augustin de Coulomb, however this can be confusing as the same letter may be used to describe temperature in Celsius. In automotive systems, the coulomb can be used to describe the charge stored in batteries or the amount of charge flowing through a circuit over time.
Farads 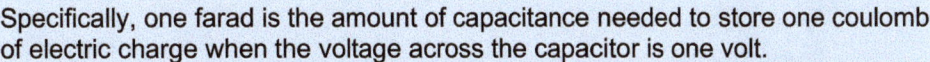	Farad, named after Michael Faraday, is a unit of electrical **capacitance**. It measures a capacitor's ability to store electrical charge. Specifically, one farad is the amount of capacitance needed to store one coulomb of electric charge when the voltage across the capacitor is one volt. Capacitance is often represented in technical information or documentation as: F - Farads, named after Michael Faraday. In automotive systems, capacitors are used in circuits for energy storage, **signal filtering**, and power stabilisation.
Henry 	Henry, named after Joseph Henry, is a unit of **inductance**. It measures the ability of an inductor (like a coil) to store energy in a magnetic field when an electrical current flows through it. Specifically, one henry is the amount of inductance needed to induce a voltage of one volt when the current through the inductor changes at a rate of one ampere per second. Inductance is often represented in technical information or documentation as: H - Henries, named after Joseph Henry. In automotive systems, the henry is useful in understanding components like ignition coils and inductors in circuits.

Electrical Essentials

Table 1.1 Units of measurement

Unit	Description
Joules 	**Joule**, named after James Prescott Joule, is a unit of energy. It measures the amount of work done or energy transferred when one **newton** of force moves an object one meter in the direction of the force. Energy is often represented in technical information or documentation as: J - Joules, named after James Prescott Joule. In automotive systems, joules can be used to describe the energy stored in batteries, the work done by motors, or the energy required to perform mechanical tasks.
Celsius 	Celsius, named after Anders Celsius, is a unit of **temperature** measurement. It is based on the scale where 0°C represents the freezing point of water, and 100°C represents the boiling point of water under standard atmospheric pressure. Temperature is sometimes represented in technical information or documentation as: °C - Celsius, named after Anders Celsius, but care must be taken not to confuse it with the letter C when used to describe electric charge. In automotive systems, it is important to understand that heat and electrical resistance are very closely linked. Heat will increase electrical resistance in a circuit, reducing current flow, and resistance will normally create heat, in most cases wasting electrical energy.
Fahrenheit 	Fahrenheit, named after Daniel Gabriel Fahrenheit, is a unit of temperature measurement. It is based on the scale where 32°F represents the freezing point of water, and 212°F represents the boiling point of water under standard atmospheric pressure. Temperature is sometimes represented in technical information or documentation as: °F - Fahrenheit, named after Daniel Gabriel Fahrenheit, and is sometimes used for temperature readings, particularly in regions where it is the preferred standard. In automotive systems, it is important to understand that heat and electrical resistance are very closely linked. Heat will increase electrical resistance in a circuit, reducing current flow, and resistance will normally create heat, in most cases wasting electrical energy.

Electrical Essentials

Electromotive Force (EMF) - the voltage generated by a power source, such as a battery or alternator, to drive electric current through a vehicle's electrical system.

Open circuit voltage (OCV) - the voltage measured across a power source when no load is connected, and no current is flowing.

Potential Difference (Pd) - the voltage difference between two points in a vehicle's electrical circuit, driving current flow.

Closed-circuit voltage (CCV) - the voltage measured across a power source when it is connected to a circuit and supplying current.

Oscillate - the repetitive variation of a signal, voltage, or mechanical movement in a vehicle system.

Resistance - the opposition to electrical current flow in a vehicle's circuit, measured in ohms (Ω).

Power - the rate/speed at which work is done.

Frequency - the rate at which an electrical or mechanical signal oscillates in a vehicle system, measured in Hertz (Hz).

Charge - the accumulation or movement of electrical energy in a vehicle's system.

Capacitance - the ability of a vehicle's electrical system or components to store and release electrical charge, measured in farads (F).

Signal filtering - the process of removing unwanted noise or interference from electrical signals in a vehicle's system to ensure accurate data transmission and sensor performance.

Inductance - the property of a circuit or component, such as a coil or motor winding, that resists changes in current flow by generating a magnetic field.

Joule - a unit of energy (J) representing the work done when one watt of power is applied for one second.

Newton - a unit of force (N) that measures the push or pull exerted on an object.

Temperature - the measurement of heat within a vehicle's systems.

Electrical Essentials

Circuit Properties and Voltage Types

The difference between EMF, Pd, OCV, CCV

Voltage, or electrical potential (pressure or force), behaves differently depending on whether an electrical circuit is switched on or off. The acronyms OCV (Open Circuit Voltage) and CCV (Closed Circuit Voltage) are often used to differentiate between these two states.

- OCV – Open Circuit Voltage (switched off).
- CCV – Closed Circuit Voltage (switched on).

An electromotive force (EMF) often represents the highest electrical pressure waiting to do some work and is mostly associated with the voltage when the circuit is switched off. Conversely, a potential difference (Pd) is the voltage measured when a circuit is switched on and current is flowing. A potential difference can be higher than an electromotive force if a circuit is under the influence of electric charge, or lower than the electromotive force if components are consuming electrical energy or the circuit is discharging.

Ground and earth

Two automotive electrical terms that are often confused or misused are 'Ground' and 'Earth.' The confusion might arise from the common association with 'planet Earth' and 'the ground'. However, in automotive electrical circuits, it is crucial to understand the distinct meanings of these terms.

- Ground represents the lowest common voltage potential on an electrical circuit. This is often accepted to be 0 volts but will depend on the circuit being tested.
- Earth describes a low-resistance electrical connection returning back to the power source in a direct current (DC) circuit or a low-resistance path away from the main circuit in alternating current (AC), often used to protect the system and operator, normally via some form of circuit breaker.

 Think of 'ground' in a DC electrical circuit as the floor of a house, where all the furniture rests at a common level—it's the reference point or base level. In contrast, 'earth' can be compared to a safety exit that directs unwanted guests (excess electrical energy) out of the house and away from harm. While 'ground' provides stability within the circuit, 'earth' ensures protection by safely diverting energy away when needed.

Electric circuits

For electrons to move from one atom to another, they need a continuous path, known as a **circuit**. This allows an electron to move forward as it is replaced by another one from behind. Without a circuit, the electrons cannot flow because the last electron in the conductor has nowhere to go. If the circuit is broken, we refer to it as losing **continuity**.

In other words, electrons need a complete loop to move. If the loop is broken, the electrons can't move because they have nowhere to go.

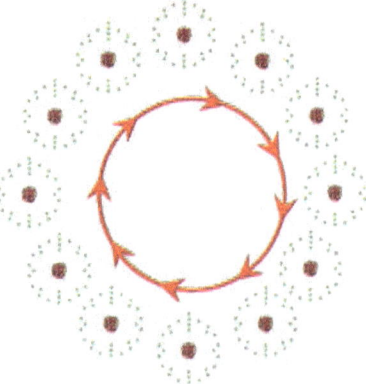

Figure 1.4 Copper atoms forming a loop

Electrical Essentials

Series and parallel circuits

Two main types of electrical circuit are used in the construction of motor vehicles:

- **Series**
- **Parallel**

Series circuit

In a series circuit, devices are connected one after another in a single line. They all share the same circuit, so they divide the electricity based on how much power each device uses. If you add more devices to the circuit, each one gets only a portion of the available voltage. The more power a device needs, the more electricity it uses.

In simpler terms, in a series circuit, all devices are lined up in a row. They share the electricity, and each device only gets a part of it. The stronger the device, the more electricity it takes.

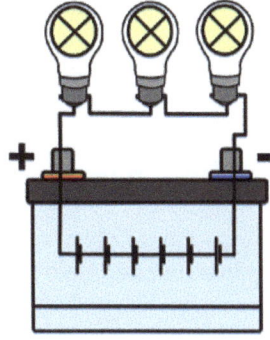

Figure 1.5 Bulbs connected in series

In a series circuit, if any one device stops working, it breaks the circuit and stops the flow of electricity. This means all the other devices in the circuit will also stop working.

Parallel circuit

In a parallel circuit, devices are connected side by side. Each device has its own power supply and return path to the supply. Because of this, all devices receive the full voltage and can operate at full power.

When you add a device to a parallel circuit, it creates a new branch or pathway. This allows more current to flow, which is inversely proportional to the resistance in that branch. This is known as a current divider.

If one device in a parallel circuit stops working, the others will continue to work. This is because each device has its own separate pathway for electricity.

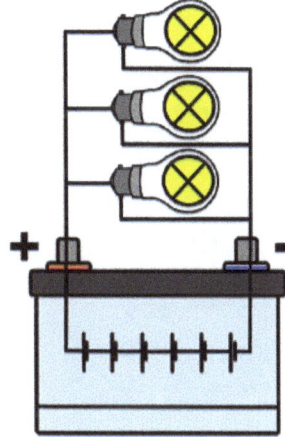

Figure 1.6 Bulbs connected in parallel

In a series circuit, the components are connected one after the other, and the current flows through each component in turn. The voltage across each component can be different and is determined by its resistance according to Ohm's law. The total resistance of a series circuit is equal to the sum of the individual resistances.

In a parallel circuit, the components are connected on different branches of the circuit and the voltage across each component is the same. The current flowing through each branch is determined by its resistance according to Ohm's law. The total resistance of a parallel circuit is less than any of the individual resistances.

Electrical Essentials

Circuit - a continuous and unbroken loop.

Continuity - the unbroken electrical path in a vehicle's circuit, ensuring current flows properly.

Series - connected one after another.

Parallel - connected side-by-side.

Electromagnetism

Electricity and magnetism are closely related, like two sides of the same coin. Both have positive and negative poles, or north and south, and both can attract and repel.

When a magnet passes a copper conductor (wire), the magnetic attraction moves electrons through the conductor, creating an electric current. Conversely, when an electric current passes through a copper conductor, it generates an invisible magnetic field. The magnetic effect of an electric current can cause movement through attraction or repulsion. This movement can be harnessed to create a motor.

Similarly, the movement of magnets past a conductor can generate an electric current, which is the principle behind a generator.

- Motors convert electrical energy into mechanical energy.
- Generators convert mechanical energy into electrical energy.

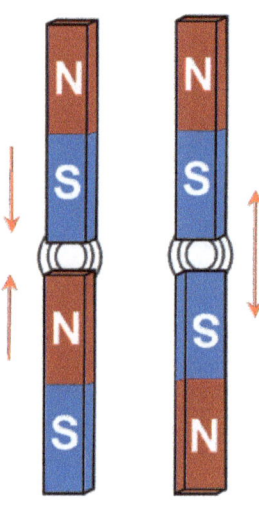

Figure 1.7 Magnets attracting and repelling

Chemical Electricity

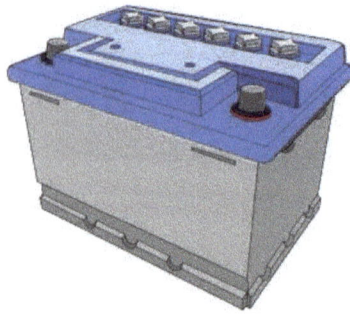

Electrical energy can be converted into chemical energy, which can be stored and transported in a battery. This process is reversible, meaning chemical energy can be turned back into electricity. Therefore, a charged battery provides a portable source of electricity that can be used as required.

Figure 1.8 A battery

While batteries are an efficient way to store and transport electrical energy, they also pose chemical hazards and must be handled with care. Batteries contain substances such as acids, heavy metals (like lead and cadmium), and other chemicals that can be harmful to humans, animals, and the environment.

Electrical Essentials

Alternating Current AC and Direct Current DC

There are two types of electric current:

- Direct Current (DC): This occurs when electrons move in a single direction within a circuit, driven by a stable potential. Simply put, DC is like a one-way street for electrons.
- Alternating Current (AC): This happens when electrons change direction periodically, driven by an oscillating potential. An AC output is sometimes referred to as a 'phase'. In simple terms, AC is like a two-way street for electrons, allowing them to move back and forth.

Alternating current (AC) voltages are often described using their RMS values. RMS stands for root-mean-square, and it is a method of measuring the effective value of AC voltage or current. This is the equivalent value of direct current (DC) voltage or current that would generate the same amount of heat or power in a resistor. Essentially, RMS value is approximately the peak value of an AC wave divided by the square root of 2, making it comparable to a DC voltage equivalent.

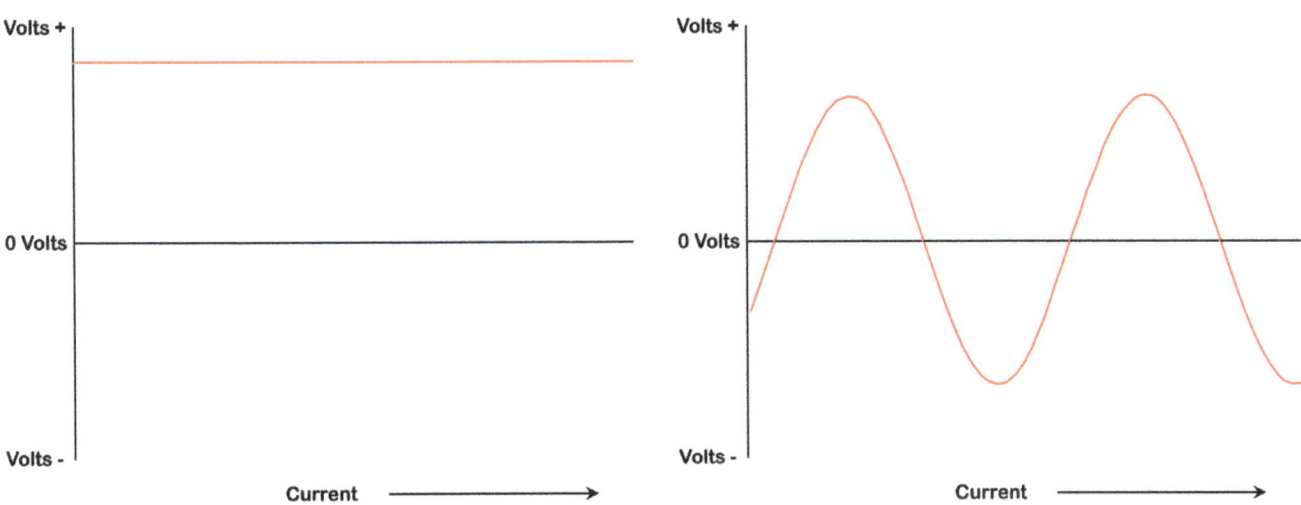

Figure 1.9 Direct current (DC)　　　　**Figure 1.10** Alternating current (AC)

Electricity, whether in the form of alternating current (AC) or direct current (DC), poses significant dangers if not handled properly. It is crucial to understand these risks to prevent accidents and ensure safety.

Both AC and DC electricity can be useful and powerful, but they must be treated with the utmost respect and caution to prevent harm to yourself and others. Always follow safety protocols and consult a professional when in doubt.

Electrical Essentials

Risks of Direct Current (DC)
- DC systems provide a stable and continuous flow of electricity, but if mishandled, they can cause severe burns or electrical shock due to the constant nature of the current.
- High-voltage DC circuits are particularly dangerous, as they lack the natural zero-crossing points found in AC, making it harder to interrupt the flow of energy during an emergency.

Risks of Alternating Current (AC)
- AC electricity changes direction periodically, which can cause severe muscle contractions, potentially making it difficult to release a live wire.
- Even at lower voltages, AC is highly capable of causing fibrillation of the heart, leading to life-threatening conditions.
- AC circuits often involve higher voltages, increasing the risk of severe injury or death from electrical shock.

General Safety Tips
- Avoid direct contact with live wires or exposed electrical components.
- Always switch off the power source before repairing or handling electrical systems.
- Use insulated tools and correct Personal Protective Equipment (PPE).
- Be aware of environmental hazards, such as water, which can amplify the dangers of electrical currents.
- Do not attempt to handle high-voltage systems without proper training and authorisation.

The automotive industry is a high-risk environment, and no matter what precautions are taken, there is always the possibility of accidents occurring that may lead to personal injury. The following advice is not a substitute for first aid training and will only give you an overview of the action you may need to take. You should take care when you attempt to administer first aid that you do not place yourself in danger. Be very careful about what you do because the wrong action can cause more harm to the casualty.

Good first aid always involves summoning appropriate help.

Electric shock: This may be caused by contact with live wires, components, or faulty equipment.
It can result in burns, muscle spasms, cardiac arrest, or death.

To treat an electric shock victim, you should:

- Turn off the power source or isolate the system if possible.

- Check for breathing and pulse and start CPR if needed.

- Seek medical attention as soon as possible.

Electrical Essentials

Understanding a direct current circuit can be tricky, so let's use a water tower analogy to explain it:

- Reservoir of Water at the Top (Battery): This is like the battery in a circuit. It stores the energy, just like the reservoir stores water.

- Pipe Leading from the Bottom of the Reservoir (Wire): This is like the wire in a circuit. It's the path that the energy (or water) travels along.

- Tap on the End of the Pipe (Switch): This is like a switch in a circuit. When you open the tap, water flows; when you close it, the water stops. Similarly, a switch controls whether electricity flows in a circuit.

- Water Wheel at the End (Motor): This is like a motor in a circuit. The flowing water turns the wheel, just as flowing electricity powers the motor.

Therefore, in this analogy, when you open the tap (switch), water (electricity) flows from the reservoir (battery) through the pipe (wire) and turns the wheel (motor). When you close the tap, everything stops.

- Quantity of Water (Current): The amount of water flowing through the pipe is like the current, or amperage, in an electric circuit.

- Water Pressure (Voltage): The pressure of the water in the pipe is similar to the voltage in an electric circuit.

- Tap Slowing Water Flow (Resistance): The way a tap slows down the flow of water is like resistance in an electric circuit.

- Water Wheel's Work Rate (Power): The speed at which the water wheel works is like the power in an electric circuit.

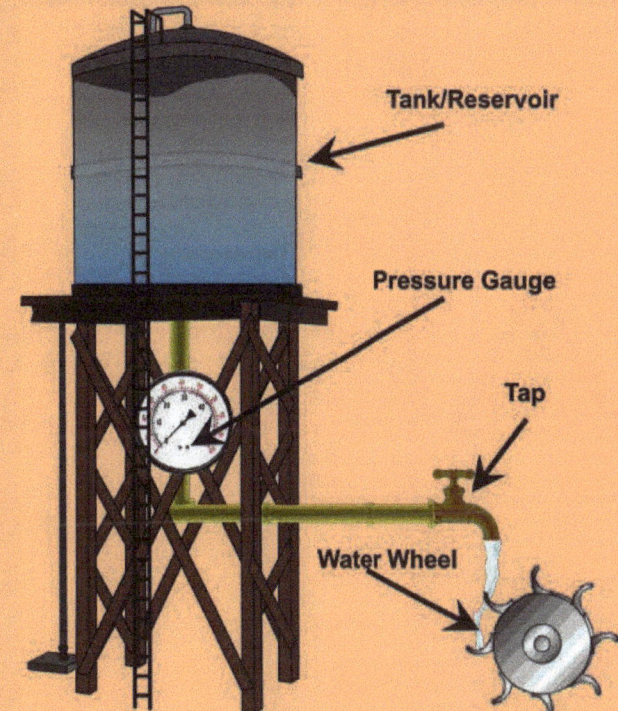

Therefore, in this analogy, the amount and pressure of water represent the current and voltage. The tap represents resistance, and how fast the water wheel works represents power.

Electrical Essentials

High and Low Voltage

Electrical voltage creates a potential danger when it comes to the possibility of electric shock or electrocution. Once the electrical pressure (voltage) reaches a point where it can overcome the natural resistance of the human body and a circuit is created with two points of contact in parallel to a power source, electric current will start to flow. The touch threshold (resistance) for dry human skin is often considered to be 50 volts, however, this value can be lower if the skin is wet, there are wounds present or the electrodes penetrate the skin. Once current starts to flow, 80 milliamps (remember that a milliamp is just 1000th of an amp) has the potential to cause injury or even death.

Consider lightly placing your hand on a sharp, upturned nail. Your skin has resistance, so as long as the pressure is light, the nail won't cause damage. However, if you increase the pressure on your hand, the nail will eventually overcome the skin's resistance and pierce it. In this analogy, the pressure is like voltage. It's the cause of any injury. But the size of the nail, which is similar to the amount of current, determines the extent of the damage. If the pressure or voltage stays below the threshold where it can overcome skin resistance, then the size of the nail or current doesn't matter. That's why warning signs say, 'Danger High Voltage' and not 'Danger High Current'.

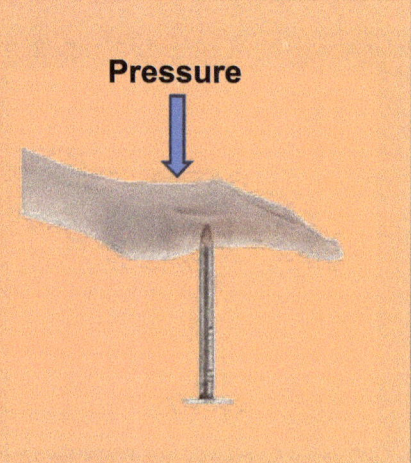

Regarding the hazards of voltage and current (amps), voltage is often considered the dangerous element, which is why warning signs typically state 'Danger High Voltage'. If a voltage exceeds the touch threshold of dry human skin, it can cause an electric current to flow, and it's this current that can cause harm. Keeping the voltage potential low reduces the risk of electric shock or electrocution. However, even with low voltage, there's still a risk of a short circuit that can lead to arcing, fire, or explosion.
Make sure you are continually evaluating the risks of electricity when conducting any diagnosis or repair.

It is also important to consider the voltages being tested during a diagnostic procedure, in order to protect the operator, equipment and vehicle.

Electrical Essentials

Ohms Law & Watts (Power) Law

Ohms law

Ohm's Law states that current flowing in a circuit is proportional to the voltage supplied and inversely proportional to the resistance.

Put simply, Ohm's Law explains how voltage, current (amps), and resistance (ohms) in a circuit are related. If you change one of these factors, it affects the others. Here's a simpler explanation using the water analogy:

1. Voltage (Pressure): If you increase the voltage in a circuit, it's like increasing the water pressure in a pipe. This makes more current flow, just like more water would flow through the pipe.

2. Resistance (Tap): If you increase the resistance in a circuit, it's like partially closing the tap on a pipe. This makes less current flow, just like less water would flow through the pipe.

Georg Ohm explained this with these formulas:

- Current (Amps) = Voltage ÷ Resistance
- Resistance (Ohms) = Voltage ÷ Current
- Voltage = Current × Resistance

Therefore, with Ohm's Law, if you know two of these measurements, you can calculate the third one. The Ohm's Law triangle is a handy tool for doing these calculations.

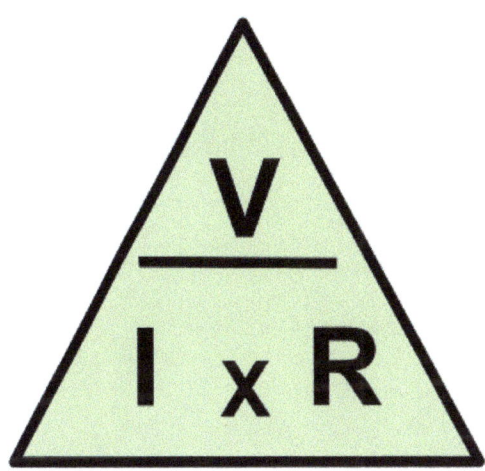

Figure 1.11 Ohms law triangle

In electrical equations, we often use the following symbols:

- **V** represents Volts.

  Sometimes, you might see the letter 'E' used instead to represent EMF (Electromotive Force), but it still means Volts.

- **I** stands for Amps.

  This letter is used to represent intensité du courant, (current intensity).

- **R** is used for Ohms.

  We use 'R' for resistance to avoid confusion with zero.

The Ohms Law triangle can help you calculate unknown units.

Here's how to use it:

Cover the unknown unit with your thumb, and the remaining letters form the calculation you need.
For example, if you don't know the amperage (I), cover the 'I' in the triangle. You'll be left with V ÷ R, which means Volts divided by resistance.

Electrical Essentials

Ohms Law, which explains the relationship between voltage, resistance, and currant (amperage), can be a useful tool for diagnosing faults in an electrical circuit. By taking measurements and comparing them using Ohms Law calculations, you can identify where the fault might be:

- Voltage (Pressure): If the voltage is lower than expected, the performance of the component might be reduced. If it's higher than expected, it could cause the component to be overworked and damaged.
- Current (Quantity/Amps): If the current is lower than expected, the component might not operate correctly. If it's higher than expected, it could mean that the component or system is being overworked.
- Resistance (Ohms): If the resistance is lower than expected, it could indicate a short circuit, where current is taking an alternative path to earth. If it's higher than expected, it could consume electrical energy and reduce system performance.

Therefore, if your voltage, current, or resistance measurements are different from what you expect, it could indicate a problem with your circuit.

Watt's (power) law

Power, measured in Watts, can be calculated similarly to Ohms Law:

- Current (Amps) = Power ÷ Voltage
- Voltage = Power ÷ Current
- Power (Watts) = Current × Voltage

You can use a power triangle, like the Ohms Law triangle, to help with these calculations.

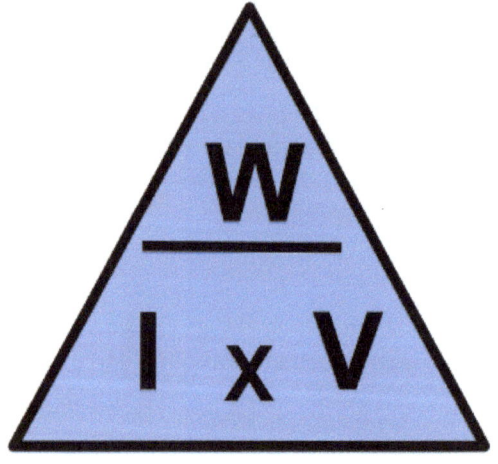

Figure 1.12 Power law triangle

- **W** represents power in Watts.

  Sometimes, you might see the letter 'P' used instead to represent power, but it still means Watts.

- **V** represents volts.

  Sometimes, you might see the letter 'E' used instead to represent EMF (electromotive Force), but it still means Volts.

- **I** stands for Amps.

  This letter is used to represent intensité du courant, (current intensity).

The Power Law triangle can help you calculate unknown units.

Here's how to use it:

Cover the unknown unit with your thumb, and the remaining letters form the calculation you need.
For example, if you don't know the amperage (I), cover the 'I' in the triangle. You'll be left with W ÷ V, which means Watts divided by Volts.

Electrical Essentials

Pulse Width Modulation and Duty Cycle

Pulse width modulation (PWM) and duty cycle, are two key terms that need to be understood when using an automotive oscilloscope.

Many electrical devices and electronic actuators can be controlled by duty cycle or pulse width modulation (PWM). These methods work by rapidly switching components on and off so that they only receive a portion of the available current or voltage, thus regulating power. Depending on the reaction time of the component being switched and the duration of power supply, variable control is achieved. This method is more efficient than using resistors to control current or voltage in a circuit. Resistors waste electrical energy as heat, whereas duty cycle and PWM operate with minimal power loss.

> Even though these terms are often interchanged, there is a subtle difference in their meaning:
>
> Pulse width modulation PWM often refers to how long something is switched on and is normally represented by a measurement of time.
>
> Although created by pulse width modulation, duty cycle, on the other hand, refers to the comparison between the amount of time something is switched on to the amount of time it is switched off (i.e., on-duty is when it's switched on; off-duty is when it's switched off). This is normally represented as a percentage.

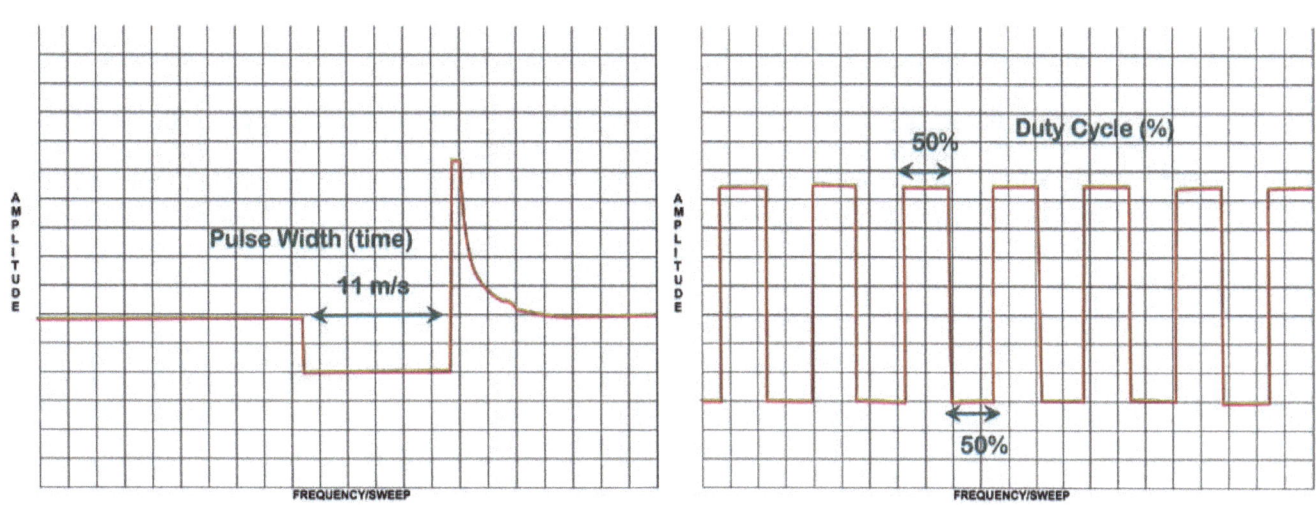

Figure 1.13 Pulse width and duty cycle

Electrical Essentials

Pull-up and Pull-down Circuits

Pull-up or pull-down circuits refer to whether a circuit is switched on by connecting it to a power source or earth.

- Connecting to the positive side of the circuit will pull the voltage 'up'.
- Alternatively, switching on the circuit by earthing to a common ground will pull the voltage 'down'.

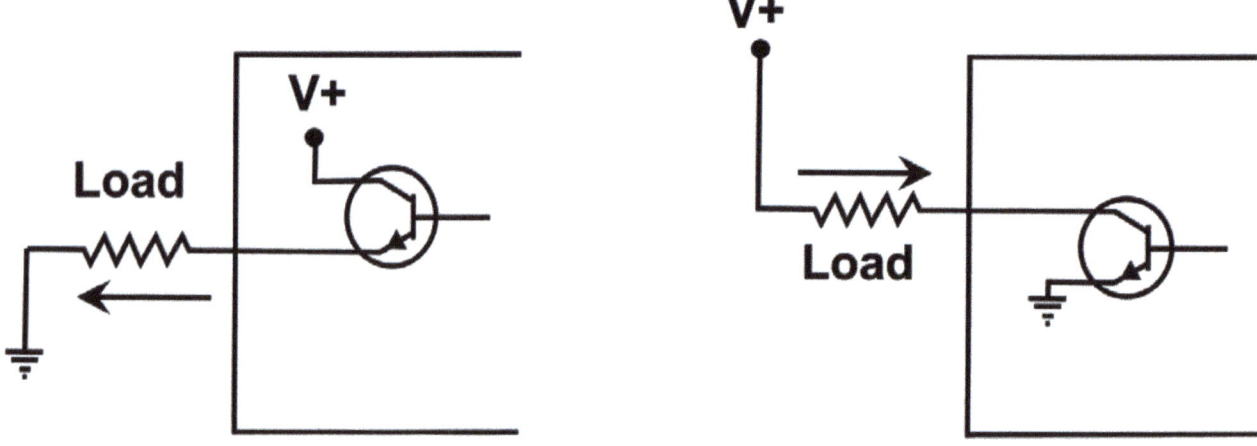

Figure 1.14 Pull-up and pull-down circuit switching

 It is often commonplace for automotive electrical control circuits to be designed as pull down.
This is because it is easier for an electronic control unit (ECU) to 'earth-out' multiple circuits or components to a common ground than provide a power supply to separate components. This is especially important when the voltage of the power supply may vary between circuit requirements.

Conclusion

Considering and understanding the importance of electrical fundamentals is key to the effective setup and use of automotive oscilloscopes. Proper knowledge of these electrical principles not only improves diagnostic accuracy but also contributes to the overall efficiency and performance of modern vehicles.

Introduction to Oscilloscopes

Chapter 2 Introduction to Oscilloscopes

This chapter will introduce you to the operation and setup of automotive oscilloscopes. It describes the physical components and provides an overview of the various functions available, explaining how these can be set or adjusted to support your diagnostic routine. The content is designed to help you become proficient quickly, with clear descriptions that allow you to develop your knowledge and understanding in a practical environment. Remember to work safely at all times and observe the relevant environmental, health and safety regulations while developing diagnostic routines that are systematic and effective.

Contents

What is an Oscilloscope	28
Quick Set-up Guide	29
Purpose and Functionality in Automotive Diagnostics	30
Types of Oscilloscopes	31
Key Features	33
Understanding Oscilloscope Components	38
Frequency/Sweep/Time Base and Amplitude	38
Probes Leads and Accessories	40
Control Panel Overview	44
Display Interface	46

The automotive industry is a high-risk environment, especially when dealing with electrical systems. The hazards of electricity are well-known but can be easily ignored due to its invisible nature. This can lead to complacency if the fundamentals of electricity are not well understood. Even with this understanding, caution is necessary. Assume that any safety systems designed for protection have failed and take precautions to minimise the risk of injury or death. Always evaluate the risks associated with any activity and implement measures to eliminate or reduce the hazards involved in any task, diagnosis, or repair. Additional risks associated with working on, or around electrical systems may include:

- Electrocution
- Strong magnetic fields
- Falling from heights
- Short circuits
- Electrical discharge/arcing
- Fire and explosion
- Chemicals

Introduction to Oscilloscopes

What is an Oscilloscope

An **oscilloscope** is an essential tool for any automotive technician working with electrical systems. Imagine it as a window into the invisible world of electricity, allowing you to see what is really happening in the circuits. It is like having x-ray vision that lets you observe the flow of electrical signals, diagnose issues, and ensure everything is functioning correctly.
At its core, an oscilloscope, or '**scope**', displays voltage changes over time on a screen.

Instead of taking a static reading and giving a direct numerical value for the result, an oscilloscope draws a picture or creates a signature of the measured results and displays this as a graphical representation known as a **waveform**. By examining these waveforms, you can understand how electrical components are functioning.

Whether you're checking sensors, ignition systems, or any other electronic part, an oscilloscope helps you pinpoint problems with precision. Having insight into an automotive electrical circuit or system through an oscilloscope increases the accuracy of fault identification, reduces frustration, and speeds up the diagnostic process.

For many years, oscilloscopes have been used in electrical engineering and medicine; however, the automotive industry also benefits from their outstanding diagnostic capabilities.

Hospitals use medical oscilloscopes to monitor heart function, and by interpreting the images produced on a screen or print-out, can make sound diagnostic decisions about the state of a person's health. They do this by measuring electrical impulses through contacts placed on the patient's body.
Similarly, an automotive oscilloscope functions like a vehicle heart monitor, with electrical contacts connected to various components on the car.

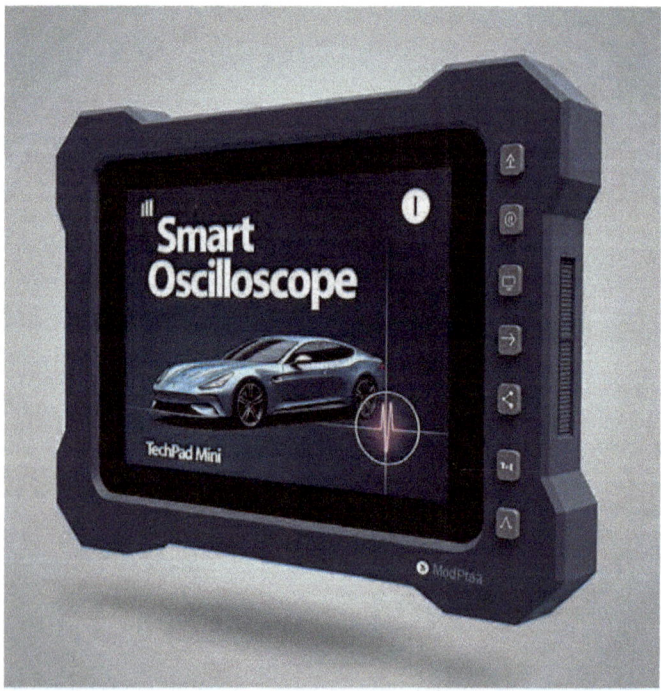

By their very nature, electrical systems and circuits are best tested while they are switched on and working. The advantage of using oscilloscopes to monitor vehicle function is that 'live' measurements can be taken while the system is running, and this extends the scope of diagnosis to test components under various operating conditions.

To break it down further, you connect **probes** to different points in an electrical system, and then the oscilloscope converts these electrical signals into visual data. You can see patterns, spikes, dips, and other anomalies that might indicate issues like **short circuits**, faulty components, or **grounding** problems.

Using an oscilloscope might seem complex, but once mastered, it becomes an invaluable part of your diagnostic toolkit. There are various types of oscilloscopes, each with distinct features; however, they all share the same fundamental purpose: making the invisible visible and helping you resolve issues efficiently.

Figure 2.1 Automotive oscilloscopes

Introduction to Oscilloscopes

However, it is worthwhile remembering that not all oscilloscopes are born equal. The capabilities and functionality can vary dramatically depending on factors such as design, size, available accessories, screen resolution, software and electronics, portability and cost. Although most scopes are capable of performing basic diagnosis, many will not have the means to conduct some of the more complex operations and activities of a superior oscilloscope.

Technicians might like the flexibility of owning more than one style of scope. A small, compact, cheap model will allow rapid access to circuit functionality and help rule in or out component or system faults quickly. A more capable oscilloscope can be used when there is the necessity for 'deep dive' or in-depth diagnosis.

If you are looking to purchase your first oscilloscope, it can be a good idea to buy a simple, cheap, and uncomplicated scope initially and get used to using it for straightforward diagnosis on a regular basis. This way, some of the issues that can arise from the perception that oscilloscopes are complicated and unnecessary can be reduced.

A cheap scope is more likely to be left out and ready for use, as its general value means it might not be treated with the same sort of reverence as one that has cost a great deal of money and investment.

If a scope is used on a regular basis, familiarity will ensure it becomes a key tool in the diagnosis of complex vehicle faults, resulting in confidence in its use and abilities.

Once your ability has overtaken the capabilities of the tool, it may then be appropriate to invest in a more expensive piece of equipment that now suits your needs.

Oscilloscope - a diagnostic tool used to visualise and measure electrical signals in vehicle systems.

Scope - in this context, an abbreviated version of the word oscilloscope.

Waveform - a visual representation of electrical signals within a vehicle's system, showing voltage or current changes over time.

Probes - specialised tools or wires used to connect an oscilloscope to a vehicle's electrical systems, ensuring accurate signal transmission for diagnosing and analysing circuits, sensors, and components.

Short circuit - an unintended electrical connection that allows current to bypass its normal path, leading to excessive current flow that can damage components or cause system malfunctions.

Grounding - the connection of a vehicle's electrical system to its metal chassis or a designated ground point.

Quick Set-up Guide

Lots of people are put off using oscilloscopes by the large box containing many wires and connectors. They feel that it will be complicated and time-consuming to set up, so they don't bother.

However, to use an oscilloscope for simple electrical testing, you only need two probes – a common and voltage wire – just like a multimeter.

To measure current, you may need an inductive clamp, and HT ignition systems will require a secondary probe. Most of the diagnostic sockets found on oscilloscopes are colour-coded, so after a quick check of the manufacturer's instructions, it should be fairly easy to know where to plug these probes in.

Introduction to Oscilloscopes

 The oscilloscope probes may come in different colours, but for the sake of simplicity we will call them red and black here.

4 step guide to connecting an oscilloscope

<u>Setup</u>

- **Step 1**: Connect the tip of the black lead to a good source of earth, such as the battery negative terminal, metal bodywork or engine. This will then only leave you with the red wire to worry about.
- **Step 2**: Now connect the red probe to the circuit to be tested.
- **Step 3**: Press the Auto Set button or adjust the scales until you see an image on the screen.
- **Step 4**: After some practice, you will become familiar with the patterns and waveforms created by different vehicle systems.

Purpose and Functionality in Automotive Diagnostics

Automotive oscilloscopes play a crucial role in identifying and diagnosing complex electrical faults that may otherwise be difficult to detect with conventional tools. By converting electrical signals into visual waveforms, oscilloscopes enable you to observe and analyse the behaviour of circuits, sensors, and actuators in real-time while the system is operating.

The primary purpose of an oscilloscope is to provide a detailed and accurate representation of electrical activities, such as voltage changes over time. This visualisation helps you understand the performance and condition of various vehicle components, ensuring that any anomalies or irregularities are promptly identified. For instance, an oscilloscope can reveal issues such as intermittent signal losses, irregular voltage spikes, improper grounding, and even some mechanical issues that can lead to malfunctions.

In terms of functionality, oscilloscopes are essential for testing and diagnosing a variety of vehicle systems. They are used to monitor the output of sensors, such as those used in engine management for oxygen, throttle position, and mass airflow, and to assess the performance of actuators like fuel injectors and ignition coils.

Additionally, oscilloscopes can be used to inspect the integrity of communication networks within the vehicle, such as CAN (Controller Area Network) and LIN (Local Interconnect Network) buses, and may even 'serial decode' messages sent and received.

As many hybrid and electric vehicles are a vital part of the world's transport network, the electric propulsion, charging, and advanced driver systems require a diagnostic tool that will allow you to interact with both low- and high-voltage systems, in order to identify issues and faults.

Introduction to Oscilloscopes

Finally, with the addition of specialist adapters and accessories, such as pressure or vibration transducers, mechanical issues can also be analysed in detail and with accuracy in a visual representation that is unavailable elsewhere.

A comprehensive understanding of oscilloscopes and their functionality is crucial for any automotive technician, as these devices provide invaluable insights into the health and operation of vehicle electrical systems. Learning to use them can significantly enhance diagnostic accuracy and efficiency, leading to more reliable and timely repairs.

Using an oscilloscope to test the systems of a hybrid or fully electric vehicle poses the risk of exposure to high-voltage systems and associated hazards. High-voltage system testing should never be carried out unless you have specific training, experience, equipment, and Personal Protective Equipment (PPE).

After any diagnostic test, all systems should be fully and correctly reassembled and tested for operation and safety.

Types of Oscilloscopes

Oscilloscopes come in various types and styles, each with its own properties, benefits, and drawbacks. Understanding these differences is crucial for selecting the right oscilloscope for your specific needs.

Handheld oscilloscopes

Handheld oscilloscopes are portable and convenient for mobility, speed, and access. They are compact and may be battery-operated or powered via the vehicle's auxiliary system, making them ideal for on-the-go diagnostics. However, their smaller display and limited functionality compared to other types can be a drawback for more complex analysis. Despite this, they are ideal for quick tests to confirm basic circuit function.

Figure 2.2 A handheld oscilloscope

Portable oscilloscopes

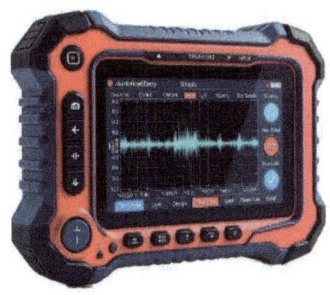

Portable oscilloscopes bridge the gap between handheld and benchtop models. They offer a balance of portability and functionality, often featuring larger displays and more advanced capabilities than handheld units. They are sometimes combined with diagnostic scan tools, either integrated into the design or attached as an accessory. This combination allows one tool to perform two tasks, such as measuring serial data and oscilloscope functions, although usually not simultaneously.

Figure 2.3 A portable oscilloscope

Introduction to Oscilloscopes

An oscilloscope can be used to test vehicle electrical systems in detail but should never be confused with the graphing meter function found on many scan tools.
A graphing meter converts live serial data into graphical representations of sensor and actuator parameters (PIDs).
The main difference is that an oscilloscope provides immediate, true readings from components in real-time, while a graphing meter presents interpreted data processed by the ECU (essentially second-hand information).

Benchtop oscilloscopes

Benchtop oscilloscopes are designed for extensive diagnostic work in a fixed location. They offer high performance, with advanced features, multiple channels, and large displays. While they provide the most comprehensive analysis for tasks such as electronic control unit (ECU) testing, their lack of portability and higher cost can be limiting factors for regular vehicle diagnostic work.

Figure 2.4 A benchtop oscilloscope

Laptop-based DSOs

Laptop-based Digital Storage Oscilloscopes (DSOs) leverage the processing power and display capabilities of laptops or PCs. They are often highly portable and adaptable, offering advanced features and the ability to store and analyse data directly on the computer. However, they depend on the availability of a laptop and may require additional software installation and configuration.

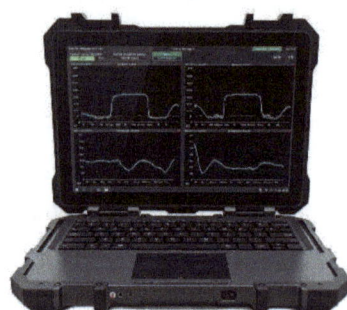

Figure 2.5 A laptop-based DSO

Analogue oscilloscopes

Analogue oscilloscopes sometimes referred to as a 'lab scope' display waveforms in real-time and are known for their simplicity and reliability. They are excellent for observing rapidly changing signals, but their lack of storage capability and limited measurement functions can be significant drawbacks in modern automotive diagnostic tasks.

Digital oscilloscopes

Digital oscilloscopes convert electrical signals into digital data for analysis, offering a wide range of advanced features such as storage, automatic measurements, and digital signal processing. They provide high accuracy and versatility, making them suitable for complex diagnostics. However, they can be more expensive and may have a steeper learning curve compared to more basic models.

Understanding the properties, benefits, and drawbacks of each type of oscilloscope helps you make informed decisions on the most suitable tool for your diagnostic needs.

Introduction to Oscilloscopes

Regardless of the complex technical capabilities and tests that can be conducted with some advanced oscilloscopes, do not be discouraged from purchasing or using them.
Start with the straightforward setup and operation; the other functions can be studied and mastered as your confidence and knowledge grow.

Key considerations for your oscilloscope:

- Power: Identify the required power source and its potential impact on any electrical interference.
- Input Voltages: Determine the maximum input voltage and ensure it aligns with the specifications of the systems and circuits you are testing.
- Software: Verify that the software is current and understand the available functionalities.

Key Features

Channels

The term 'channels' when used in association with an oscilloscope refers to the individual pathways through which electrical signals are input and displayed on the screen. Each channel can be thought of as an independent signal path, allowing you to observe multiple signals simultaneously.
Imagine having several separate voltmeters combined in one diagnostic tool, which can display their results independently on the same screen for comparison.
This is particularly useful in automotive diagnostics, where different components and systems may need to be monitored simultaneously.
A channel is a dedicated input on the oscilloscope used to connect an electrical signal for measurement and display. Each channel corresponds to a separate trace or waveform on the oscilloscope screen, enabling the comparison of multiple signals.

Setup

Step 1: Connect the probes or leads to the points in the circuit where the signals need to be measured. (This will require access to **breakout** or **back-probe** locations.)

Step 2: Select the appropriate channel on the oscilloscope's control panel or menu.

Step 3: Adjust the settings (such as **frequency**, **sweep**, time-base, and **amplitude**) to ensure the waveform is displayed at the correct scale.

Step 4: Use the oscilloscope's interface to switch between channels or display multiple channels simultaneously.

Introduction to Oscilloscopes

Automotive oscilloscopes typically have two or more channels, allowing you to conduct a comparative analysis of different signals.

For instance, you could measure the voltage signal from an engine sensor on one channel and the effect it has on a fuel injector on another, making it easier to diagnose issues by comparing the two signals side by side.

Channels are fundamental in automotive diagnostics for various applications:

- Simultaneously monitoring multiple sensors and actuators.
- Comparing input and output signals to identify faults or discrepancies.
- Observing the interaction between different systems, such as the ignition and fuel systems.
- Capturing and analysing complex waveforms that occur simultaneously in different parts of the vehicle's electrical system.

A 'floating ground-referenced' oscilloscope can be necessary when measuring voltages that are not referenced to earth ground or when working with circuits where the ground potential is not stable or known.

It is important to understand that some multi-channel oscilloscopes share a common ground internally (i.e. the earth connection for each channel grounds to one shared point within the scope itself). This provides a common point of reference for measurements and provides a protected earth.
This no real problem for straightforward **differential voltage measurements**.

However, some sensors may have their own signal ground, to reduce the possibility of electrical **interference** from other circuits or systems. If an oscilloscope with a common ground is used for testing this style of sensor, it may have an effect on the operation of the component and vehicle.

In this case an oscilloscope with an isolated or 'floating input' on the channels is needed or connection via a separate component called a **'differential probe'** is required.

Breakout - a diagnostic connector that attaches to the vehicle cable or wiring harness, allowing easy access to circuit signals for testing, monitoring, and troubleshooting without damaging connectors.

Back-probe - a method of attaching diagnostic test probes in the back of a socket or plug to make contact with an electrical connector or terminal without damaging wiring or insulation.

Frequency - the rate at which an electrical or mechanical signal oscillates within a vehicle system, typically measured in hertz (Hz).

Sweep - the movement of the oscilloscope's trace across the display, capturing and visualising electrical signals over time.

Introduction to Oscilloscopes

Amplitude - the peak value of an electrical signal's voltage or current in a vehicle system.

Differential voltage measurements - measuring the voltage difference between two specific points in a vehicle's electrical circuit.

Interference - the disruption of a vehicle's electrical signals caused by unwanted electromagnetic noise or currents, which can affect the performance of electronic components and systems.

Differential probe - a tool used with oscilloscopes to measure the voltage difference between two points in a vehicle's electrical system, offering accurate analysis while minimising interference and noise.

Sampling rate

The sampling rate is a crucial factor in the operation of automotive oscilloscopes, determining how often the oscilloscope captures data points from the signals it measures. This rate is usually expressed in samples per second (S/s) or mega samples per second (MS/s). Understanding the sampling rate is essential requirement to ensure accurate diagnostics and effective troubleshooting.

Sampling rate, also known as sample rate or sample frequency, refers to the number of times per second an oscilloscope samples the signal it measures. Higher sampling rates generate more data points, offering a more detailed and accurate representation of the signal. For instance, an oscilloscope with a sampling rate of 1 MS/s (Mega Samples per second) captures one million samples per second.

A high sampling rate ensures a finer **resolution** of the captured signal. Low sampling rates may miss rapid changes or intricate details, which can result in inaccurate analysis and potential misdiagnosis. Essentially, the higher the sampling rate, the better the resolution and the more details that can be observed.

The general rule is that the sampling rate must be at least twice the highest frequency present in a signal to accurately reconstruct the waveform. Failure to meet this requirement can result in **aliasing**, a phenomenon where higher frequencies appear as lower ones, distorting the measurement.
In automotive applications, signals from sensors and electronic control units (ECUs) often contain high-frequency components. A high sampling rate is essential to accurately capture these components, enabling effective diagnosis of malfunctions, intermittent events, and signal integrity issues.

Resolution - the smallest voltage or time increment that an oscilloscope can accurately distinguish.

Aliasing - a signal distortion caused when the sampling rate of the oscilloscope is too low to accurately capture high-frequency signals, resulting in misleading or incorrect waveforms.

Introduction to Oscilloscopes

Making effective use of sampling rate

To make the most of an appropriate sampling rate, you should consider the following aspects:

- Ensure that the oscilloscope's sampling rate is suitable for the frequency range of the signal being measured. For instance, if diagnosing a signal with frequencies up to 100 MHz, the oscilloscope should have a sampling rate of at least 200 MS/s (Mega Samples per second).
- Many oscilloscopes allow adjustment of the sampling rate. you should familiarise yourself with these settings to tailor the sampling rate to specific diagnostic requirements. Reducing the sampling rate can be beneficial for low-frequency signals, conserving memory and processing power.
- In situations with noisy signals or when capturing low-frequency components, averaging multiple samples can improve signal clarity. By combining several samples, the oscilloscope can reduce the impact of random noise, enhancing the accuracy of the measurement. Sometimes known as **'resolution enhancement'**, this can often be found in the channel settings of a DSO.
- Transient events, such as spikes or sudden drops in voltage, require high sampling rates to be accurately captured. You should select an oscilloscope with a higher sampling rate to effectively monitor and analyse these rapid changes in automotive systems.

Sample rate applications in automotive diagnostics

- Sampling rate is crucial when diagnosing electronic control units (ECUs) that process high-frequency signals from various sensors. Accurate sampling ensures that inconsistencies in sensor data are detected and rectified efficiently.
- Anti-Lock Braking System (ABS) sensors can generate high-frequency signals to monitor wheel speed. A high sampling rate oscilloscope is essential to capture and diagnose issues related to ABS functionality accurately.
- The rapid pulses in ignition systems require oscilloscopes with good sampling rates to measure and analyse ignition timing and performance. Accurate sampling helps identify misfires or irregularities in ignition patterns.
- Vehicles that use high-speed communication **protocols** like CAN, LIN, and FlexRay need an appropriate sampling rate to analyse signal integrity, detect data transmission errors, and ensure protocol compliance.

It is important to understand that the sampling rate of many automotive oscilloscopes reduces as more channels are used. For example, an oscilloscope specified at of 400 MS/s may achieve only 200 MS/s when two channels are used or 100 MS/s if four channels are used.
If the Digital Storage Oscilloscope (DSO) is operated via a laptop or PC, the sampling rate might also be limited by the USB cable streaming speed.

Resolution enhancement - a technique used with oscilloscopes to improve the clarity and detail of captured waveforms by reducing noise and refining signal precision, aiding accurate diagnostics.

Protocols - standardised communication methods that enable data exchange between vehicle electronic systems.

Introduction to Oscilloscopes

Bandwidth

Bandwidth refers to the range of frequencies an oscilloscope can accurately measure. It is a critical parameter that defines the highest frequency signal that the oscilloscope can capture with a reasonable accuracy.

In automotive diagnostics, oscilloscopes must possess sufficient bandwidth to capture the high-frequency signals generated by various electronic components and systems. For instance, Electronic Control Units (ECUs), Anti-Lock Braking Systems (ABS), ignition systems, and communication networks such as CAN, LIN, and FlexRay often operate at frequencies that demand high-bandwidth oscilloscopes.

Choosing an oscilloscope with appropriate bandwidth is essential for accurate signal representation. If the bandwidth is too low, the oscilloscope may fail to capture the true nature of the signal, leading to inaccurate diagnostics. For example, a 100 MHz oscilloscope is generally capable of accurately capturing signals up to 20 MHz. A general rule of thumb is that the bandwidth should be at least five times higher than the frequency of the signal being measured.

When selecting an oscilloscope for automotive applications, you should consider the following:

- System Requirements: Identify the highest frequency signals in the systems being diagnosed and ensure the oscilloscope's bandwidth exceeds this frequency.
- Future-Proofing: Anticipate future diagnostic needs and choose an oscilloscope with higher bandwidth to include advances in automotive electronics.
- Bandwidth Limitations: Be aware of the oscilloscope's bandwidth limitations and avoid measuring signals that exceed its capacity to ensure accurate readings.

Display resolution

Display resolution refers to the number of data points or **pixels** that an oscilloscope can display on its screen and is often expressed in terms of vertical and horizontal resolution.

Vertical resolution, measured in **bits**, determines the oscilloscope's ability to distinguish between different signal amplitudes. A higher vertical resolution provides more precise amplitude measurements, which is particularly important for capturing small signal variations in complex systems.

Horizontal resolution, on the other hand, affects the oscilloscope's ability to display detailed waveforms over time. This resolution is influenced by the oscilloscope's sampling rate - the number of samples it takes per second. A higher sampling rate results in better horizontal resolution, allowing you to observe intricate signal changes and diagnose issues with greater accuracy.

Factors influencing display resolution include:

- Sampling Rate: A higher sampling rate improves horizontal resolution, capturing more details in the signal over time.
- Memory: Adequate memory allows the oscilloscope to store more data points, enhancing resolution during long-duration captures.
- Vertical Resolution: Higher bits in vertical resolution provide finer **granularity** for amplitude measurements, crucial for distinguishing small signal variations.
- Display Quality: A high-quality screen with sufficient pixels ensures clear visualisation of waveforms and measurement data.
- Zoom Functionality: The ability to zoom in on specific waveform sections aids in detailed examination and accurate diagnostics.

Introduction to Oscilloscopes

Pixels - the individual points of light that make up the graphical display on the oscilloscope screen.

Bits - the number of levels an oscilloscope's analogue-to-digital converter (ADC) can divide a signal into.

Granularity - the level of detail in signal representation. Higher granularity allows for finer distinctions in voltage levels and more precise signal analysis.

Understanding Oscilloscope Components

Display

A high-quality screen with sufficient pixels ensures clear visualisation of waveforms and measurement data. The display quality directly impacts the clarity with which you can observe and interpret the captured signals. A screen with high resolution allows for detailed examination, making it easier to identify and analyse signal anomalies.

The ability to zoom in on specific sections of a waveform is essential for detailed examination and accurate diagnostics. Zoom functionality enables you to focus on particular areas of interest, providing a closer look at signal characteristics, and improving analysis.

The display interface, including the grid, waveforms, and measurement data, is where you visualise the captured signals. Understanding how to read and interpret the display interface is crucial for accurate diagnostics and analysis.

Probes and leads

Probes and leads are the physical connections between the oscilloscope and the circuit being tested. They come in various types and are designed for different uses. Proper handling and selection of probes and leads are crucial for obtaining accurate measurements and avoiding signal interference.

Control panel

The control panel of an oscilloscope consists of various knobs, buttons, and menus that allow the user to adjust settings such as time base, volts per division (volts/div), and other measurement parameters. Understanding the control panel of your own oscilloscope is essential for effectively operating the scope and obtaining precise readings.

Frequency Sweep Time Base and Amplitude

Frequency, sweep, time base, and amplitude are essential parameters in understanding and using oscilloscopes effectively, particularly in automotive diagnostics. They describe how measurements are laid out on the 'x' and 'y' axes of a graph.

Introduction to Oscilloscopes

Frequency

Frequency refers to the number of times a waveform repeats itself within a specific period along the 'x' axis of the display. In diagnostics, accurately measuring frequency is crucial for evaluating the performance of components such as ignition systems, fuel injectors, and sensors. Understanding frequency helps in diagnosing misfires, irregularities in sensor signals, and other performance issues.

Sweep

The term 'sweep' comes from an older form of oscilloscope and its process of moving an electron beam horizontally across the display screen of a **cathode-ray tube**, which allowed for the visual representation of signals. By adjusting the sweep speed, you can capture fast or slow events with the required level of detail. On a more modern oscilloscope, it is often used as a term to describe how quickly the **trace** moves across the screen.

Time base

The time base setting on an oscilloscope controls the horizontal sweep speed or the amount of time represented by each division on the display screen graph. This setting is vital for matching the oscilloscope display with the signal being measured. Proper adjustment of the time-base setting allows you to view complete **cycles** of waveforms, making it easier to analyse signal behaviour over time.

Adjusting the time base will allow you to stretch or shrink the width of the waveform, allowing you to focus on specific areas. This process is often referred to as **'zooming'**.
Some oscilloscopes offer a magnifier function which allows an area on the screen to be selected for closer inspection. When the capture is paused, the magnifier can be used to zoom in and examine areas of interest.

Amplitude

Amplitude measures the magnitude or height of the waveform, representing the strength of the signal. In the context of an oscilloscope, amplitude is often displayed in volts per division (volts/div). Properly setting and interpreting the amplitude is essential for understanding the signal's power and diagnosing issues related to voltage levels in electrical systems.

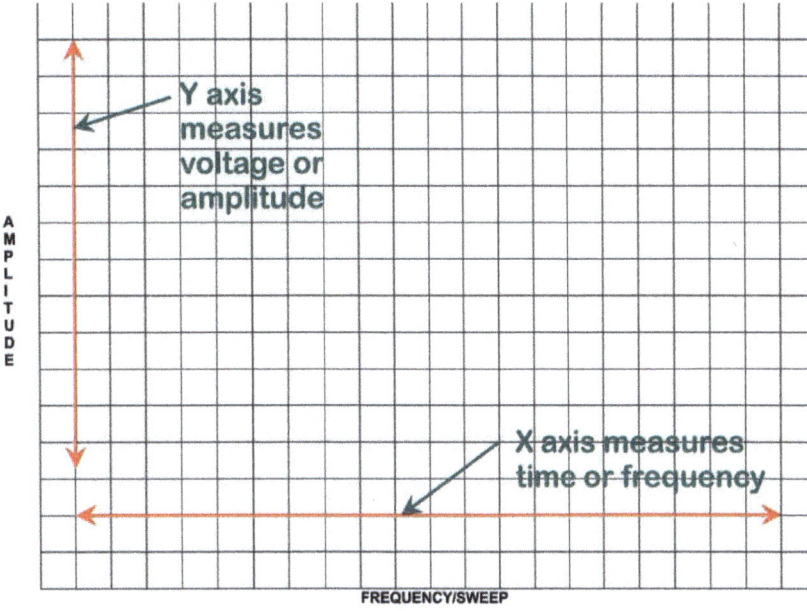

Figure 2.6 An oscilloscope screen

Introduction to Oscilloscopes

Think of a graph like a ladder leaning against a wall:

- The x-axis is the ground, running horizontally. It represents a stable foundation, just like the surface you stand on.
- The y-axis is the ladder, reaching vertically. It climbs upward, much like values increasing on a graph.

So, when remembering axes, just picture the ground beneath your feet (x-axis) and the ladder stretching toward the sky (y-axis).

An easy way to remember which axis is which on a graph is to simply say, 'X is across' (a cross).

If you are unsure of the most appropriate voltage or timescale to use for a particular test, simply select a value somewhere in the middle and then move up and down the scale until the pattern is displayed to your satisfaction.

Frequency - the rate at which an electrical or signal waveform oscillates in a vehicle's system, measured in Hertz (Hz).

Cathode-ray-tube - an older display technology which operates by directing electron beams onto a phosphor-coated screen to produce images.

Trace - a recorded waveform or data output that represents the electrical or signal activity within a vehicle system, typically captured using an oscilloscope or diagnostic tool for analysis and troubleshooting.

Cycles - the repeated operational phases within a vehicle system.

Zooming - a method of enhancing the view of a waveform by magnifying specific sections of the signal.

Probes Leads and Accessories

To take measurements, the oscilloscope will often be provided with a series of accessories which help you access the various system components. The more comprehensive the scope, the more accessories are often available, and this will mean that a large case full of wires and connectors is provided. Unfortunately, this promotes the misconception that the equipment will be difficult to set-up and use.

Many accessories will be test leads, probes and clips to help you connect the scope to the desired component/circuit, however, some have specialist functions and are designed specifically for purpose. Examples of some diagnostic accessories can be seen in **Table 2.1**.

Introduction to Oscilloscopes

Table 2.1 Oscilloscope diagnostic accessories

Accessory	Description
Probes and back probes 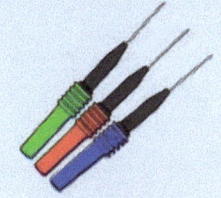	Many oscilloscopes have a selection of interchangeable probes with different styles of contact point. Some probes may have a very thin point that can be used to '**back-probe**' a connection by pushing it into the rear of the components' electrical connectors so it can touch the terminal and test the signal without causing damage to the wiring or circuit.
Crocodile clips 	Crocodile or alligator clips are small clamps, often manufactured with teeth on the inner jaws, which can be fitted to the end of the test leads and connected to the circuit or component to provide hands-free operation.
Voltage test leads 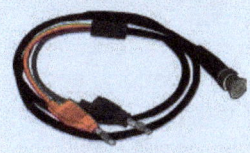	Voltage test leads are the most common type of lead used with oscilloscopes. They are designed to measure voltage signals within the circuit. They attach to the scope using a **bayonet** style connector known as a **BNC**. The test end of the leads will normally have a connector that can be accessorised with probes or crocodile clips.
Passive probes 	Passive probes do not contain any active electronic components. They are typically more robust and less expensive than active probes. Passive probes are suitable for general-purpose measurements and are often used for lower-frequency applications.
Differential probes 	Differential probes are used to measure the voltage difference between two points in a circuit. These probes are essential for applications where the signals are not referenced to ground and are useful in minimising the risk of **ground loops** and other **interference**.
Secondary ignition connectors 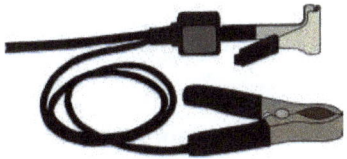	In order to test the high voltage produced by a secondary ignition system, adapters are available which clip around the **high-tension** lead and detect the pulse created when a spark is produced, converting it to a waveform for display. As many modern ignition systems no longer have high-tension leads, it may not be possible to use these connectors without an additional extension lead adapter which fits between the ignition coil and the spark plug.

Introduction to Oscilloscopes

Table 2.1 Oscilloscope diagnostic accessories

Accessory	Description
Inductive secondary ignition probes 	If an ignition system is designed as **coil-on-plug** (COP), some scope manufacturers provide secondary ignition probes that are **inductive**. These probes vary in shape and design but can simply be held against the high-tension component being tested and produce a waveform for display.
Fused circuit loop adapters 	This is a small loop of wire which is connected in series to the fuse box and can then be used to create a location where an inductive amp clamp can take readings from the circuit. Although the fuse has been removed, there is an in-line fuse holder in the loop itself, which is used to protect the circuit while testing is conducted.
Inductive amps measurement clamps 	An inductive amp clamp is a vital accessory if you want to measure current using your oscilloscope. Unlike a standard ammeter, which must be connected in series with a circuit, an inductive clamp simply clips around a wire of the circuit that you want to test, the circuit is switched on, and the readings taken. The inductive clamp will have its own internal power source (battery) and, in many cases, uses the **Hall effect** principle to sense the magnetic field created around a wire when the circuit is switched on. The strength of the magnetic field is then converted into a waveform displayed on the oscilloscope screen and is interpreted as amperage.
Attenuators 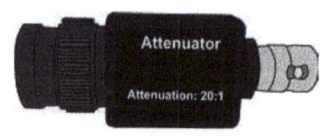	An **attenuator** is a resistor component which is connected in series with the test lead and probe. It is designed to allow the oscilloscope to measure higher voltage values than those calibrated on the readout screen. They will normally be designed to reduce the displayed amplitude in multiples of ten (i.e. x10, x20 etc.) and this will be marked on the casing of the attenuator. If an attenuator is used during your testing routine, you must multiply the voltage readings displayed on the graph by the value shown on the attenuator.
Keyless entry detector 	A diagnostic probe used to detect the carrier signal frequency from the vehicle keyless entry system. They are normally a short range detector to allow the detection of individual pickup coils.
Ultrasonic parking sensor detector 	A diagnostic probe used to detect the correct operation of **ultrasonic** parking sensors. It is a form of microphone designed to operate in the frequency rage used by the most common styles of parking sensor.

Introduction to Oscilloscopes

Table 2.1 Oscilloscope diagnostic accessories

Accessory	Description
Pressure transducers 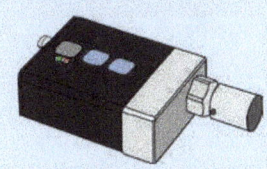	A pressure **transducer** is an accessory which, with a set of adapters, can be connected to vehicle systems or components and record physical measurements of pressure for either fluid, gas or vacuum. When pressure is sensed, it is converted into an electrical signal that can be interpreted as an amplitude on the display of the oscilloscope.
Noise, vibration, and harshness (NVH) probes 	NVH probes are used to measure noise, vibration, and harshness levels within the vehicle. These probes help in identifying and diagnosing issues related to engine performance, suspension systems, and overall vehicle comfort.

 Many inductive amp clamps will have a zero button which should be used to calibrate the probe.

Back-probe - an electrical connection made at the component plug which contacts the terminals without damage to the wiring's insulation.

Bayonet - a quick-release electrical connector featuring pins that fit into corresponding slots and twist-lock securely into place, ensuring a reliable connection while allowing easy disconnection.

BNC - BNC (Bayonet Neill–Concelman) connector is a coaxial connector with a bayonet-lock mechanism, widely used in testing and diagnostic equipment to transmit signals with minimal interference and ensure a secure, quick connection.

Ground loop - multiple ground paths with differing electrical potentials (voltages), leading to unwanted current flow and signal interference, often causing noise in audio or electronic systems within a vehicle.

Interference - unwanted electrical noise or distortions that affect signal clarity during waveform analysis.

High-tension - the high-voltage electrical energy used in ignition systems to generate sparks in combustion engines.

Coil-on-plug - an ignition system where each spark plug has its own coil mounted directly on top.

Introduction to Oscilloscopes

Inductive - a process where a voltage is created inside a conductor by magnetism.

Hall effect - the principle where a magnetic field influences electrical voltage in a sensor.

Attenuator - a component accessory used to reduce the strength of a signal recorded for testing.

Ultrasonic - the use of high-frequency sound waves for sensing and detection in vehicles.

Transducer - a device that converts one form of energy to another: pressure into an electrical signal for example.

Handling and care

Where possible:

- Always store probes and connectors in their designated cases or holders to prevent damage and tangling of wires. Keep them in a dry, dust-free environment to avoid corrosion and deterioration.
- Clean the probes regularly with a soft cloth and isopropyl alcohol to remove dirt and oil. Avoid using harsh chemicals or abrasive materials that can damage the probe's surface or connectors.
- Regularly inspect the probes for signs of wear and tear, such as frayed cables, damaged connectors, or broken tips. Replace any damaged probes immediately to maintain measurement accuracy and safety.
- Handle the probes with care to avoid bending or twisting the cables excessively. Use strain reliefs and cable management accessories to prevent stress on the probe connections.
- Calibration: Periodically calibrate the probes according to the manufacturer's instructions to ensure their accuracy. Calibration helps maintain the integrity of the measurements and extends the lifespan of the probes.
- Connection: When connecting probes to the oscilloscope and the circuit under test, ensure that the connections are secure and free from corrosion/oxidation or debris. Use appropriate adapters and connectors to match the probe type and measurement requirements.

Proper handling and care of probes and connectors are crucial for obtaining accurate and reliable measurements with your automotive oscilloscope.

Many probes and accessories will be interchangeable with different oscilloscopes due to the BNC connection type or use of adapters. Care should be taken, however, to ensure any readings are calibrated to the oscilloscope being used, as the probes or accessories may have design features which make their scaling unsuitable or different from those shown on the display.

Some oscilloscopes have the capability for the user to scale the output by creating 'custom probes' (*see Chapter 6*).

Control Panel Overview

Understanding the various knobs, buttons, switches, and menus is essential for effectively using an oscilloscope. **Table 2.2** provides a general overview of controls to help you get started, however, as each scope design will vary, practice with your own equipment to learn the specific location and function.

Introduction to Oscilloscopes

Table 2.2 Controls

Control	Function
Time base	The time base controls the horizontal scale of the oscilloscope display, representing the time duration of the signal being observed. Adjusting this sets the time per division (e.g., milliseconds per division - ms/div), allowing you to focus in or out on the waveform.
Volts/Div	The volts per division (v/div) setting controls the vertical scale of the oscilloscope display, representing the signal amplitude. Adjusting this changes the voltage per division, enabling you to amplify or reduce the signal's magnitude (height) on the screen.
Trigger level	The trigger level sets the voltage level at which the oscilloscope begins to capture the signal. This helps stabilise a repetitive waveform and provides a clear view of the signal's behaviour.
Auto set	The auto setup is a convenient feature that automatically adjusts the oscilloscope's settings to provide a stable and readable display of the input signal. This is particularly useful for quickly capturing and viewing signals without manual adjustments. It normally sets the amplitude and time base derived from the activity in the circuit being tested. This can provide a starting point, from where fine adjustments can be made by the operator.
Run/Stop	The run/stop button controls the acquisition of the signal. Pressing this button starts or stops the signal capture, allowing you to freeze the waveform for more detailed analysis.
Single trigger/capture	The single trigger button captures a single waveform and then stops. This is useful for capturing intermittent events or single-shot signals.
Magnifier/zoom	Some oscilloscopes offer a magnifier function which allows an area on the screen to be selected for closer inspection.

Menus

The main menu is the starting point for accessing all of the oscilloscope's features. It typically includes options for setting the time base, voltage scale, and trigger settings, among other parameters. **Table 2.3** provides a general overview of menus to help you get started, however, as each scope design will vary, practice with your own equipment to learn the specific location and function.

Table 2.3 Menus

Menu	Function
Trigger	The trigger menu allows you to configure the triggering settings in detail. You can select the trigger source, mode (e.g., edge rise or fall), and level. Proper trigger settings are crucial for capturing stable and accurate waveforms.
Measurement	Measurement menus provide options for measuring various signal parameters, such as frequency, amplitude, peak-to-peak voltage, and duty cycle. These measurements are displayed on the screen and can be used for detailed analysis.
Display	Display menus allow you to customise the appearance of the oscilloscope's screen. You can adjust the grid style, waveform colour, and intensity, as well as enable or disable cursors and rulers for precise measurements.
Save/Recall	The save/recall menu enables you to save waveform data and oscilloscope settings for future reference. This is useful for documenting measurements and quickly recalling specific configurations.

Introduction to Oscilloscopes

 Although not every oscilloscope may have all of these functions, understanding the basic configuration of your oscilloscope's knobs, buttons, and menus is essential for effectively diagnosing and analysing automotive electrical signals.

Display Interface

Understanding the display interface of your oscilloscope is vital for accurately diagnosing and analysing electrical signals. The key components of the display are the grid, waveforms, and measurement data.

The grid

The grid, also known as a **graticule**, is the foundational element of the oscilloscope screen, providing a reference framework for visualising waveforms and measurement data. The grid typically consists of horizontal and vertical lines that create a graph-like appearance. Each line represents a specific unit of time or voltage, helping you to quantify and compare signal characteristics.

Some oscilloscopes allow you to adjust the grid style which can enhance visibility and precision:

- Horizontal Lines: These lines represent voltage levels, allowing you to see the amplitude of the waveform.
- Vertical Lines: These lines represent time intervals, helping you to measure the duration of signal events.

Waveforms

Waveforms are graphical representations or signatures of electrical signals. They are displayed on the oscilloscope screen and show how the voltage of a signal changes over time.

Understanding waveforms is essential for diagnosing issues in automotive electrical systems:

- Sinewave: Represents a smooth, periodic oscillation, common in alternating current (AC) circuits and some sensor outputs.

- Square Wave: Consists of sharp transitions between high and low voltage levels, typically found in digital signals.

- Triangle Wave: Displays a linear rise and fall in voltage, useful for analysing ramp signals.

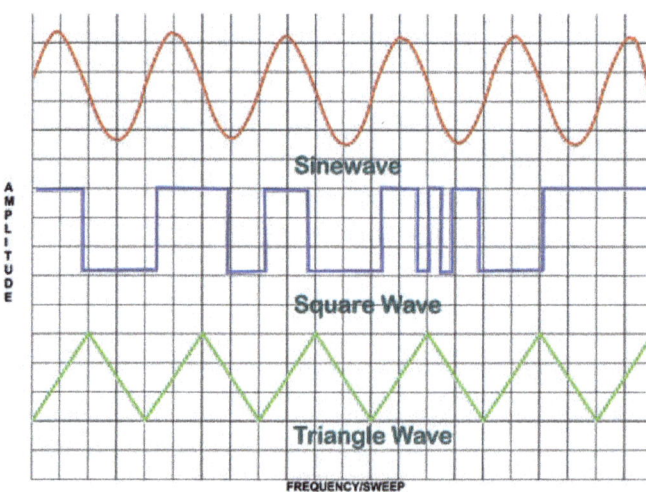

Figure 2.7 Waveform shapes

Introduction to Oscilloscopes

 Analysing the shape, frequency, amplitude, pulse width and duty cycle of waveforms can help identify abnormalities and pinpoint faults.

More information on waveforms and their meanings can be found in the companion book: *Automotive Oscilloscopes Waveform Analysis*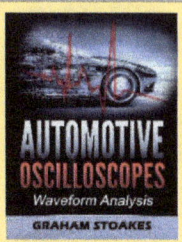

Measurement data

Measurement data is displayed alongside the waveforms, offering precise numerical values for various signal parameters. It can be enhanced by adding measurement points on the screen. This data can be crucial for detailed analysis and comparison:

- Frequency: The number of **cycles** the waveform completes in one second, measured in Hertz (Hz).
- Amplitude: The maximum voltage level reached by the waveform, indicating the signal strength.
- Peak-to-Peak Voltage: The difference between the highest and lowest voltage levels of the waveform.
- Duty Cycle: The percentage of one period in which the signal is active (which may be high or low depending on the circuit design).

 Many oscilloscopes provide on-screen **cursors** and **rulers** that can be used to measure these parameters directly from the waveforms. Also, some models allow users to save and recall measurement data for future reference, facilitating documentation and the construction of a waveform library.

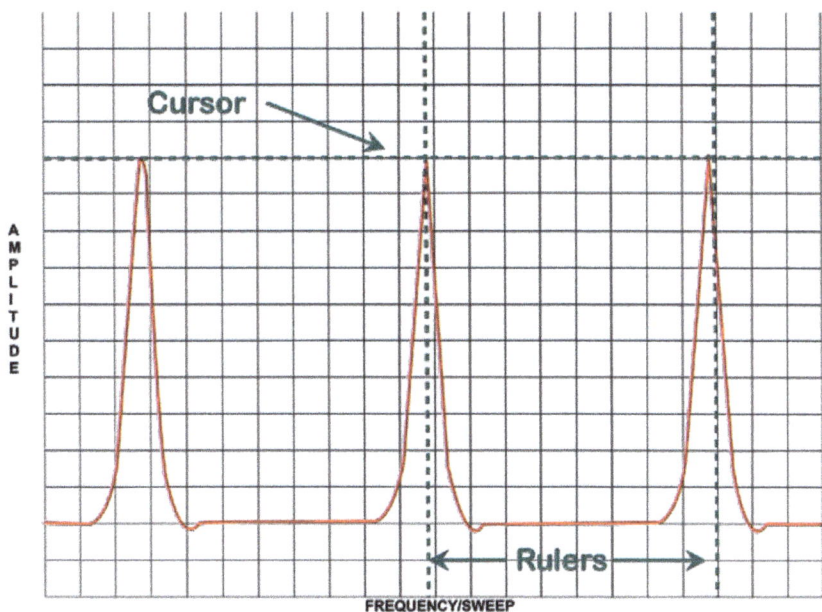

Figure 2.8 Cursors and rulers

Introduction to Oscilloscopes

- Regularly calibrate your oscilloscope and probes to ensure accurate measurements.
- Familiarise yourself with the customisation options for the grid and waveforms to optimise visibility.
- Use cursors and rulers to make precise measurements directly on the display.
- Document and save key measurements for future reference and comparison.
- Save your favourite settings for quick access.
- Create your own personal waveform library.

Graticule - the grid overlay on the oscilloscope screen, consisting of horizontal and vertical lines.

Cycles - patterns that represent specific events or operations within a vehicle's electrical system, such as ignition timing or sensor signals. Each cycle typically includes a period and amplitude that reflect the system's behaviour over time.

Cursors - adjustable markers on the screen used to measure specific points on a waveform, such as time intervals, voltage levels, or frequency, enabling precise analysis of automotive signals.

Rulers - fixed reference lines on the screen that help measure waveform parameters like time and voltage, providing a visual guide for accurate analysis of signal characteristics.

Conclusion

Oscilloscopes are essential for diagnosing and repairing automotive electrical systems. Your skill with these tools will improve your ability to address complex issues and enhance techniques, which boost your expertise and service quality.

Setup and Use

Chapter 3 Setup and Use

In this chapter, you will learn how to set up and use an automotive oscilloscope for diagnosing vehicle issues and faults. The process begins with understanding your device's initial setup and configuration. Correctly connecting probes to the oscilloscope, mastering probing techniques to avoid damaging circuits, and calibrating the device for precise measurements are crucial steps you must follow. The chapter will also cover various power supply options for your oscilloscope, ensuring it is always ready for use - whether through battery, mains, or USB power. By the end of this chapter, you will be equipped with the foundational knowledge necessary to set up and use your automotive oscilloscope efficiently, paving the way for accurate diagnostics and smoother automotive repairs.

Contents

Initial Setup	**50**
Power Supply Setup	**52**
Basic Configurations	**53**
Practical Usage	**66**
Common Automotive Applications	**70**
Troubleshooting and Maintenance	**85**
Safety Considerations	**87**

The automotive industry is a high-risk environment, especially when dealing with electrical systems. The hazards of electricity are well-known but can be easily ignored due to its invisible nature. This can lead to complacency if the fundamentals of electricity are not well understood. Even with this understanding, caution is necessary. Assume that any safety systems designed for protection have failed and take precautions to minimise the risk of injury or death. Always evaluate the risks associated with any activity and implement measures to eliminate or reduce the hazards involved in any task, diagnosis, or repair. Additional risks associated with working on, or around electrical systems may include:

- Electrocution
- Strong magnetic fields
- Falling from heights
- Short circuits
- Electrical discharge/arcing
- Fire and explosion
- Chemicals

Setup and Use

Initial Setup

Oscilloscopes come in a variety of designs and styles. Though the following descriptions are generic and should cover most types, you may need to practice on your own oscilloscope. More detailed descriptions of settings can be found later in this chapter.

The oscilloscope should be connected to a suitable power source and switched on.
Choose and enable the required number of channels for your diagnostic routine. Select and configure the input channels on your oscilloscope. Proper channel setup allows for better signal analysis and comparison.

Depending on the circuit(s) or component(s) being tested, select suitable probes, connectors, or accessories. (*See **Table 2.1** Chapter 2*). Ensure that the probes are securely connected to the oscilloscope to prevent any loss of signal or inaccurate readings. The connections should be firm, and it is important to use the appropriate connectors for your specific oscilloscope model.

Make sure you are aware of the voltage potentials in the circuits being tested, and take any necessary precautions to protect yourself, the equipment, and the vehicle. For example, you may need to use an attenuator, a differential probe, or high-voltage PPE.

Perform any calibration required to ensure that your test results are accurate. This often requires following the manufacturer's instructions. Alternatively, you can perform a basic calibration by connecting to a known good voltage source, such as the 12-volt auxiliary battery, and verifying the results. These results can then be compared or checked using a digital voltmeter.

Setup

Step 1
- Set up basic configurations to meet your specific needs.

Step 2
- If your oscilloscope supports multi-screen display, adjust the settings accordingly.

Step 3
- Configure the sampling and buffer settings to capture and store data effectively for accurate waveform analysis.

Step 4
- Determine whether the circuit is AC or DC and make your selection. (This can be switched later if needed).

Step 5
- If the expected voltage values and timescales are uncertain, start with a high value and adjust as necessary during testing.

Step 6
- Connect the probes securely to the circuit or components being tested, following proper probing techniques.

Setup and Use

Step 7
- Some oscilloscopes offer guided testing or pre-sets, which are preprogrammed settings and information for specific system, actuator, or sensor testing. These can be helpful for getting you in the general area but may require fine-tuning once testing has commenced.

Step 8
- After following all necessary health and safety precautions, operate the system and adjust the oscilloscope settings to ensure the best waveform sample is achieved. If your oscilloscope has an auto-setting button, this can often be used to quickly achieve an initial automated setting for both amplitude and time base. Fine adjustments can then be made to further refine your results.

Step 9
- To manually adjust the time base and voltage division (volts/div), up and down, left and right, or plus (+) and minus (–) buttons are often provided. By using these buttons, you can stretch or shrink the waveform pattern until it fits the screen, optimising the display for the signals you are observing. This helps in getting a clear and detailed view of the waveforms.

Don't worry if you are unsure of which way to adjust the settings (up/down – left/right), just try it in small increments, and if wrong, simply go the other way.

Once a waveform is acquired, triggers can be used to stabilise the pattern, the capture can be paused, and measurements added to assess its operation.

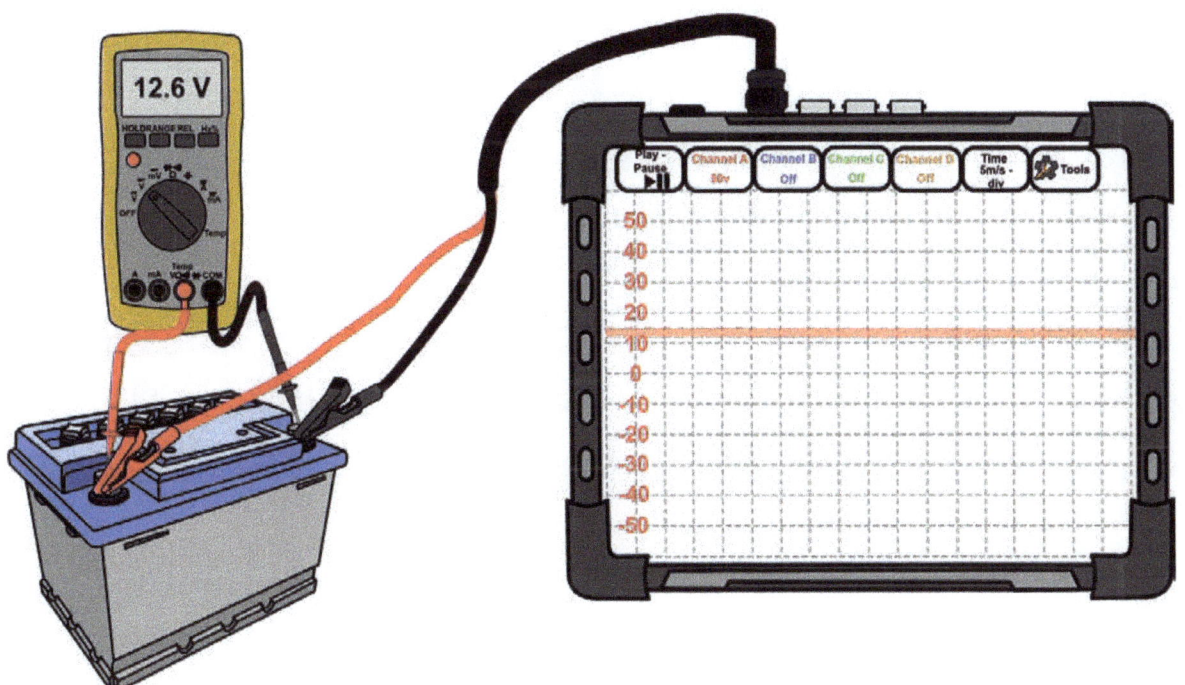

Figure 3.1 Calibrating the oscilloscope on a known good voltage supply

Setup and Use

Power Supply Setup

Battery power

A battery-powered oscilloscope provides the benefit of portability, making it ideal for roadside or outdoor work and scenarios where access to mains power is limited. One significant advantage of battery power is the elimination of **electrical noise** from power supplies that can interfere with sensitive measurements. However, the primary disadvantage is the limited operational time, which depends on battery capacity and the oscilloscope's power consumption. It is crucial to ensure the battery is fully charged before starting any tests or diagnosis.

 When using a battery powered oscilloscope, access to a power bank, portable charger or bringing a spare battery can be a practical precaution.

 **Electrical noise** - unwanted, irregular signals or disturbances within the waveform, often caused by electromagnetic interference or faulty components, which can obscure accurate diagnosis of the vehicle's electrical systems.

Mains power

Oscilloscopes powered by mains electricity are suitable for extended periods of use, such as in a workshop or garage environment. They provide a constant power supply, which is useful for prolonged testing without the need to worry about battery life. The downside is the potential introduction of electrical noise from the mains supply, which can affect the accuracy of measurements. Using a power conditioner or isolation transformer can help reduce this issue. Additionally, mains-powered oscilloscopes tend to be less portable, restricting their use to areas with power outlets.

 If extension leads are used, they should be fully unwound to avoid heat created by flowing current and induced magnetic fields which could lead to interference, damage or a fire.
Care must be taken to avoid trailing cables over walkways or areas where they might present a trip hazard.

USB power

USB-powered oscilloscopes offer a convenient and versatile option, especially when working with modern digital interfaces. They can draw power directly from a computer or a USB power bank, which increases their portability and ease of use in various settings. The primary advantage is the integration with computer software, enabling advanced data analysis and storage capabilities. However, the power supplied by USB is generally lower compared to mains or dedicated battery power, which may limit the performance and functionality of the oscilloscope. Precautions should be taken to ensure that the USB source can provide a stable and sufficient power supply to avoid interruptions during testing.

Setup and Use

Basic Configurations

Views

Some oscilloscopes provide the option of changing how the display screen is set out.

Setup

Step 1: Connect the probe(s) to the desired input channel(s).

Step 2: Select the channel(s) on the oscilloscope.

Step 3: Adjust the time base and voltage settings to obtain a clear view of the waveform.

Single screen view

The single screen view is the most basic and commonly used display mode. It shows the waveform from a single channel on the oscilloscope screen. This view is straightforward and simple, making it easy to focus on one signal at a time.

Advantages

- Simplicity and ease of use.
- Ideal for observing a single signal in detail.

Disadvantages

- Limited to one signal at a time, which may not provide a comprehensive view of the system's behaviour.

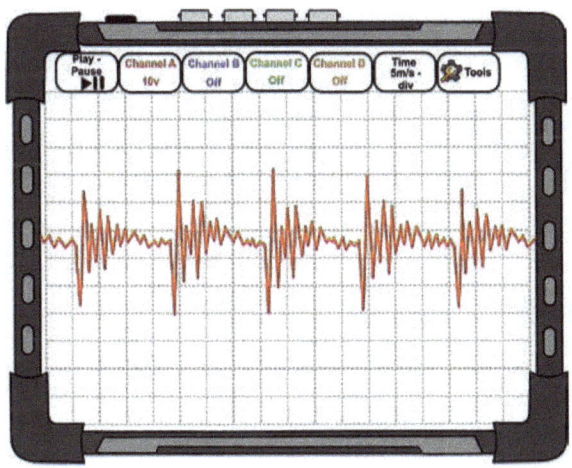

Figure 3.2 Single screen view

Setup and Use

Dual screen view

The dual screen view allows you to observe two waveforms simultaneously, each from different channels. This is particularly useful for comparing signals or monitoring the interaction between two components.

Advantages	Disadvantages
• Enables comparison of two signals at the same time, assisting in identifying correlations or inconsistencies. • Useful for analysing component interactions between components.	• Can be more complex to set up and interpret. • Screen space is divided, which may reduce the resolution of each waveform.

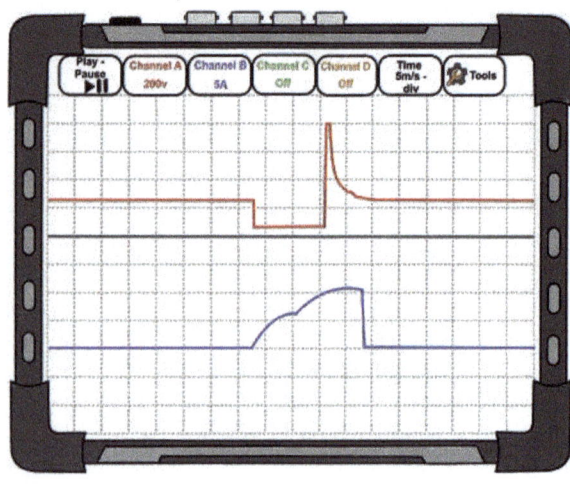

Figure 3.3 Dual screen view

Multi-screen view

For more advanced diagnostics, oscilloscopes can display multiple waveforms from several channels. This view is essential for complex systems where multiple signals need to be monitored simultaneously.

Advantages	Disadvantages
• Provides a comprehensive overview of multiple signals. • Essential for complex diagnostics and system-wide analysis.	• Increased complexity in setup and interpretation. • Each waveform may appear smaller due to the divided screen space.

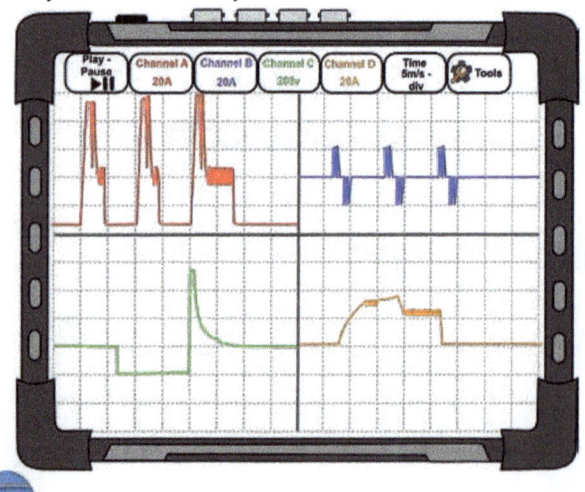

Figure 3.4 Multi-screen view

Setup and Use

Overlay view

The overlay view stacks multiple waveforms on top of each other on the same screen. This is useful for directly comparing signal shapes and timing relationships.

Setup

Step 1
- Connect the probe(s) to the desired input channel(s).

Step 2
- Enable any overlay function on the oscilloscope. (It may be possible to simply align the waveforms by adjusting their position or dragging them on top of each other).

Step 3
- Adjust the time base and voltage settings to ensure clear visibility of all waveforms.

Advantages

- Direct comparison of waveforms, ideal for timing or inconsistency analysis.
- Allows for easy identification of signal deviations.

Disadvantages

- Overlapping waveforms can cause visual clutter, making it harder to differentiate between signals.
- Requires careful adjustment to avoid misinterpretation.

Figure 3.5 Overlay view

Setup and Use

Split-screen view

The split-screen view divides the oscilloscope display into separate sections, each showing a different waveform. This is beneficial for organising the display and reducing visual clutter.

Setup

Step 1
- Connect the probe(s) to the desired input channel(s).

Step 2
- Enable the split-screen function (if available) on the oscilloscope.

Step 3
- Configure the time base and voltage settings for each section.

Advantages

- Organised display, reducing visual clutter.
- Allows for easy focus on individual waveforms.

Disadvantages

- May reduce the size of each waveform, potentially impacting resolution.
- More complex setup process.

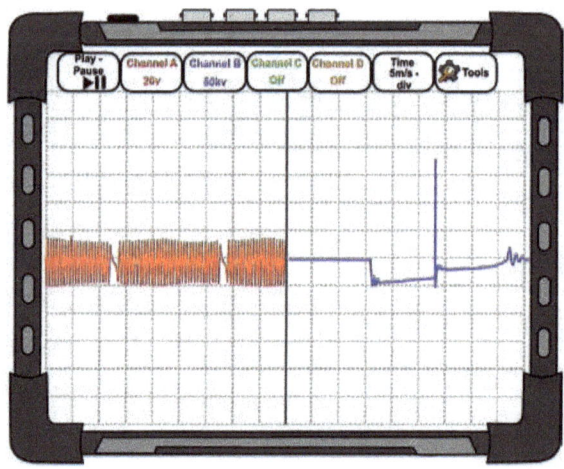

Figure 3.6 Split screen view

Setup and Use

Buffers

Buffers are an essential feature in automotive oscilloscopes, often used to temporarily store data before it is processed or displayed. This section explores the significance of buffers, their setup process, and the associated advantages and disadvantages.

What are buffers?

Buffers are the memory storage areas that capture and retain the signal data prior to its analysis and display. This temporary storage allows the oscilloscope to manage incoming data at varying speeds, ensuring that no information is lost during rapid signal changes.

Think of a buffer in an oscilloscope like a waiting room at a doctor's surgery.
Imagine the electrical signals are like patients arriving at the clinic. The doctor (oscilloscope's processor) can't see all the patients at once, so the waiting room (buffer) holds them until the doctor is ready.
This waiting room ensures patients don't get lost or mixed up—it keeps everything organised so the doctor can see each one in turn.
In an oscilloscope, the buffer stores signals temporarily, making sure they are held safely before they get processed and displayed on the screen. This is especially useful when signals arrive too quickly for the oscilloscope to handle immediately.
So, just like a waiting room ensures every patient gets seen properly, a buffer ensures every signal is captured accurately for analysis.

Importance of buffers

Buffers are crucial for effective signal analysis, particularly when dealing with high-speed or complex signals. They provide a stable environment where signal data can be processed sequentially (one after another), which is especially important for capturing transient/intermittent events or irregular patterns that may otherwise be missed.

Setup

Step 1
- Navigate to the buffer settings in your oscilloscope's menu.

Step 2
- Determine the buffer size based on the signal complexity and the desired resolution. A larger buffer can capture more data, which is useful for detailed analysis.

Step 3
- Configure any triggering options to ensure that the buffer captures the relevant parts of the signal.

Setup and Use

Advantages	Disadvantages
• Enhanced Data Capture: Buffers allow for comprehensive data capture, even during rapid signal changes. • Accurate Analysis: By storing data temporarily, buffers enable more precise and thorough analysis. • Reliable Signal Processing: Buffers manage the flow of data, ensuring that the oscilloscope processes it in an orderly manner. • Flexibility: Adjustable buffer sizes and settings provide flexibility to cater to various diagnostic scenarios.	• Increased Memory Usage: Larger buffers consume more memory, which may limit the oscilloscope's overall storage capacity. • Complexity: Setting up buffers can be complex and may require a thorough understanding of the signal characteristics and oscilloscope capabilities. • Potential Delays: The process of storing and retrieving data from buffers may introduce slight delays in signal display.

The maximum number of buffers will depend on how many samples are collected in each waveform and will be limited by the oscilloscope's memory.

Channel setup

Selecting and configuring input channels is a critical step in using your scope effectively. It involves choosing the right channels to monitor specific signals and configuring them to ensure an accurate and meaningful data capture.

Selecting input channels

Different channels are used to capture signals from various components such as sensors, actuators, and circuits. You must decide which signals are most important to the diagnostic task at hand. This involves understanding the vehicle's electrical system and identifying the signals that provide the most relevant information.

Configuring input channels

Once the appropriate input channels have been selected, the next step is configuration. Configuring input channels ensures that the oscilloscope captures and displays signals in a manner that is useful for your analysis. This includes setting the correct range and coupling options. For instance, selecting between AC and DC coupling depends on the nature of the signal being measured. Additionally, input **impedance** settings should be adjusted to match the characteristics of the signal source, preventing loading effects that could distort the signal. For example, the **back EMF** produced when a fuel injector switches off may require the use of an **attenuator**.

Impedance primarily affects AC (alternating current) because it includes resistance, inductance, and capacitance, which change how AC signals behave over time. However, DC (direct current) is only affected by resistance—since inductance and capacitance do not react to steady DC flow in the same way they do with AC.

Setup and Use

Channel synchronisation

Synchronising channels is another key part of setting up your oscilloscope. When you sync multiple input channels, you can capture different signals at the same time. This helps you see how they interact with each other. This is especially useful in automotive diagnostics, where understanding how various signals relate can help identify complex issues.

Calibration and testing

Finally, **calibration** and testing are essential to make sure the configured channels are working correctly. Regularly checking the calibration of the oscilloscope channels and probes is needed to keep them accurate. You should also perform test measurements to check that the setup captures the expected signals correctly, ensuring dependable diagnostics. For example, different channels may have different sampling rates, which could wrongly show an **edge matching** discrepancy when comparing **CAN BUS** signals between CAN High and CAN Low.

Although some oscilloscopes may not allow calibration adjustments or resets, ensuring that output data is accurately aligned improves the assessment of test results. Calibration can often be verified by using a known good input and comparing the values obtained with those from a multimeter.

Impedance - the opposition a circuit or component presents to the flow of alternating current (AC) in a vehicle's electrical system. It combines resistance, inductance, and capacitance, affecting signal integrity, power transfer, and diagnostics in automotive electronics.

Back EMF - back EMF (electromotive force) refers to the voltage generated in an electrical circuit, such as a motor or coil, when the current flow is interrupted. This counter-voltage opposes the initial current, causing a spike in the waveform.

Attenuator - a device or function that reduces the amplitude of incoming signals, allowing the oscilloscope to measure high-voltage signals accurately without damaging its circuitry.

Synchronising - the process of aligning signals, components, or systems within a vehicle to ensure proper timing and coordination.

Calibration - the process of adjusting an oscilloscope to ensure accurate signal measurement. It involves setting voltage scales, time bases, and probe compensation to maintain precise waveform analysis in vehicle diagnostics.

Edge matching - the alignment of signal edges (rising and falling) to ensure proper synchronisation and data integrity between components in a vehicle's CAN BUS communication system.

CAN BUS - a high-speed communication protocol that enables data exchange between electronic control units (ECUs) in a vehicle.

Setup and Use

Time and voltage settings

Before diving into the specifics of adjusting the time base and volts/div settings, it is essential to understand what these terms mean and how they impact the functionality of an automotive oscilloscope.

- Time Base: This setting determines how much time is represented horizontally across the oscilloscope screen. It is usually measured in seconds, milliseconds, microseconds or nanoseconds per division (sec/div, ms/div, µs/div or ns/div). Adjusting the time base allows you to zoom in or out on the waveform, making it easier to observe slow or fast events in detail.
- Volts/Div: This setting controls the vertical scale of the waveform displayed on the screen. It is measured in volts per division (v/div). By adjusting the volts/div, you can change the amplitude of the signal on the screen, making high or low voltage signals more visible and easier to analyse.

Even though essentially all amplitude settings on an oscilloscope represent voltage, the interpretation can be converted into other values such as, amperage or pressure for example, depending on the type of probe or adapter connected to the system.
The oscilloscope may have the facility to change the display measurements to match the types of reading expected, by selecting the probe type from a menu. However, in other circumstances, the conversion will need to be conducted manually.
If an attenuator is used, the amplitude may also need to be manually calculated. For example, a x20 attenuator will require the measured results to be multiplied by twenty.

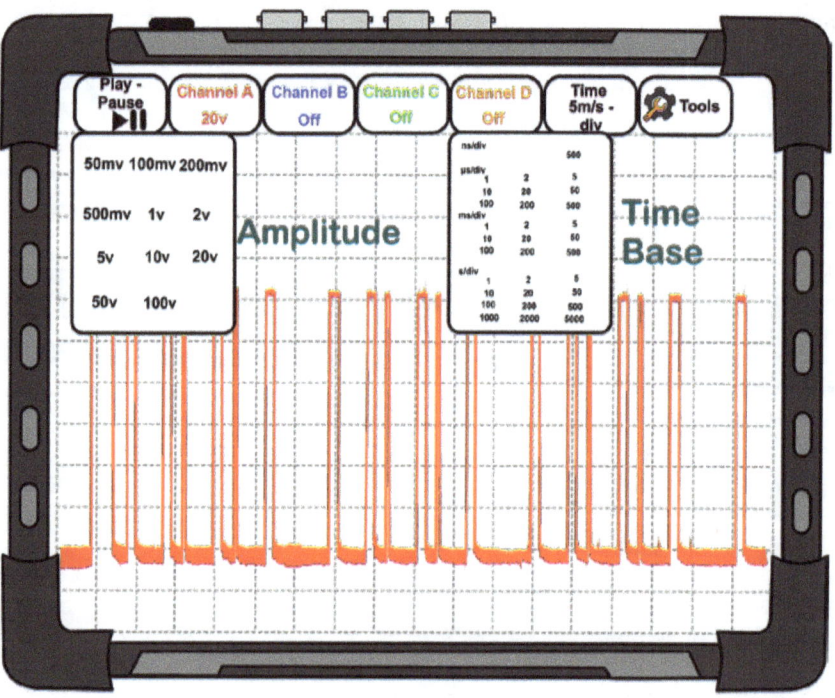

Figure 3.7 Adjusting time base and amplitude

Setup and Use

Adjusting time base

To achieve an optimal display, the time base setting must be adjusted to match the frequency of the signal being measured.

Setup

Step 1
- Identify the Signal Frequency: If possible, determine the frequency of the signal you are analysing. This might be from previous experience or available in the manufacturer specifications.

Step 2
- Set an Initial Time Base: Start with a time base setting that allows a broad view of the signal. For example, a slow time base will show more of the available capture on the screen. This might seem overcrowded at first but can provide a good starting point.

Step 3
- Fine-Tune the Time Base: Adjust the time base to stretch out the capture, focusing in on specific parts of the waveform. If the signal appears too compressed, reduce the time base (go faster). Conversely, if the signal is too spread out, increase the time base (go slower).

Step 4
- Observe Signal Details: Ensure that the time base setting allows you to observe the waveform's relevant features, such as peaks, troughs, and transitions. The goal is to have a clear and detailed representation of the signal for accurate analysis.

Adjusting Volts/Div

Properly adjusting the volts/div setting is crucial for accurately capturing the amplitude of the signal.

Setup

Step 1
- Identify the Signal Amplitude: Knowing the expected voltage range of the signal is essential. In most circumstances these are often in the lower voltage ranges of automotive auxiliary systems of 12 to 24 volts, however, remember that coils used inside solenoids etc, will produce voltage spikes (back EMF) when switched off. Care must be taken not to exceed the oscilloscope's input voltage, and some testing may require the use of an attenuator.

Step 2
- Set an Initial Volts/Div: Start with a volts/div setting that allows the signal to fit within the vertical limits of the screen. Slowly reduce the voltage until a waveform appears.

Step 3
- Fine-Tune the Volts/Div: Adjust the volts/div setting to ensure the signal occupies the majority of the screen without clipping. If the signal is too small, decrease the volts/div (bring the voltage down). If the signal is too large and exceeds the screen limits, increase the volts/div (move the voltage up).

Step 4
- Ensure Signal Visibility: The objective is to have a waveform that is neither too small nor too large, allowing you to observe its amplitude changes accurately. Proper adjustment of the volts/div setting enhances the visibility of signal details, aiding in precise analysis.

Setup and Use

Generally, oscilloscopes are not designed to be directly connected to any high voltage systems in electric or hybrid vehicles.
Any electrical test equipment used to diagnose high voltage systems must be safety categorised at a minimum of CATIII 1000V, including probes, cables, and connectors. Work on the systems of electric or hybrid vehicles should only be undertaken by individuals who have received proper training and are equipped with approved high voltage Personal Protective Equipment (PPE).

Use Cursors and Grids: Many oscilloscopes have built-in cursors and grids that help you measure and analyse waveforms. Use these features to assist in setting time base and volts/div accurately.
- Avoid Overloading: Ensure that the signal does not overload the oscilloscope input, as this can distort the waveform, lead to inaccurate measurements and may be dangerous.
- Check for Noise: Minimise noise interference by using proper grounding techniques and ensuring good connections. Excessive noise can obscure signal details and complicate analysis.
- Regular Calibration: Periodically check the calibration of your oscilloscope and probes to maintain accuracy in time base and volts/div settings.

Rulers and cursors

Rulers and cursors are essential tools for precise waveform measurements and analysis. These features help accurately determine signal parameters such as amplitude, frequency, rise time, and time intervals between events.

What are rulers and cursors?

Rulers and cursors are visual aids displayed on the oscilloscope screen that allow you to set reference points and measure specific aspects of a waveform. Rulers are typically horizontal or vertical lines that can be positioned at any point on the screen to act as fixed reference markers. Cursors, on the other hand, are movable lines or crosshairs that you can drag across the waveform to select and measure various points of interest.

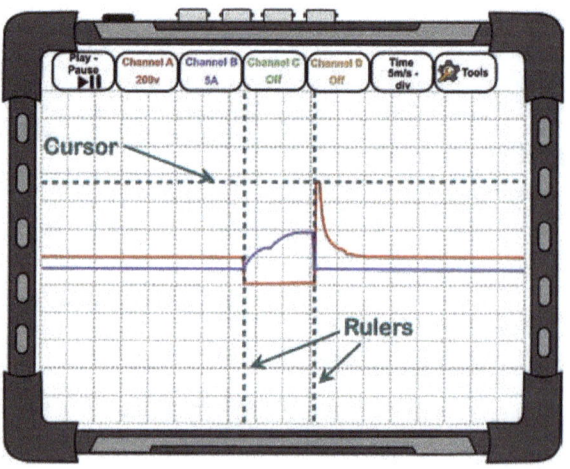

Figure 3.8 Rulers and cursors

Setup and Use

 Depending on the oscilloscope design and operating instructions, the terms 'cursors' and 'rulers' are sometimes interchanged. This is often due to the particular function they have been assigned or how they are used.

How rulers and cursors work

- Rulers: Rulers are fixed lines that provide a visual reference for measuring voltage levels (displayed as horizontal lines across the screen) or time intervals (displayed as vertical lines from top to bottom of the screen). By aligning the rulers with specific points on the waveform, you can obtain precise measurements of signal characteristics.
- Cursors: Cursors are movable lines that can be positioned at various points on the waveform. Typically, two cursors are available, allowing users to measure the difference between two points (known as the Delta Δ). For instance, placing one cursor at the beginning of a waveform pulse and the other at the end allows you to measure the pulse width. The oscilloscope's display will then show the time difference and voltage difference between the two cursors, providing valuable data for analysis.

 Think of an automotive oscilloscope like a magnifying glass over a blueprint:
- Rulers act like grid lines on the blueprint, helping you measure distances and dimensions with fixed reference points.
- Cursors are like a movable pointer, allowing you to precisely check specific details, like distances between signals or voltage levels.

Together, they help pinpoint exact values in a waveform, ensuring accurate diagnostics.

Setting rulers and cursors

Setup and use

Step 1
- Activate the Ruler/Cursor Function: Some oscilloscopes have these displayed on the screen ready for activation and some may need you to access the oscilloscope menu and enable the ruler or cursor feature. This can usually be done through the 'Measure' or 'Cursor' menu.

Step 2
- Select the Appropriate Rulers/Cursors: Choose on the vertical axis for voltage measurements and the horizontal axis for time measurements. Select the type of cursor needed (e.g., amplitude, time interval).

Step 3
- Position the Rulers/Cursors: Use the oscilloscope's control buttons or touchscreen interface (drag and drop) to move the rulers or cursors to the desired positions on the waveform. Align the rulers with specific voltage levels or time points and place the cursors at key points of interest for measurement.

Step 4
- Read the Measurements: Once the rulers and cursors are positioned, read the measurement values displayed on the oscilloscope screen. These values will indicate the voltage difference, time difference, or other relevant parameters between the selected points.

Setup and Use

Start with a Clear Signal: Ensure that the waveform is stable and free of noise before setting rulers and cursors. This will improve the accuracy of your measurements.
Use Multiple Cursors: Take advantage of multiple cursors to measure different aspects of the waveform simultaneously. For example, measure both the rise time and fall time of a pulse by placing cursors at the start and end of each transition.
Combine Rulers and Cursors: Use rulers in conjunction with cursors to obtain more comprehensive measurements. For instance, position vertical axis rulers at key voltage levels and use cursors to measure the time intervals between these levels.
Document Your Findings: Record the measurements obtained using rulers and cursors for future reference and analysis. This documentation is valuable for diagnosing recurring issues or comparing signal parameters over time.

Trigger settings

The **trigger** is an important part of an oscilloscope, which helps capture and show waveforms. It sets the condition when the oscilloscope starts collecting data, which helps you to keep the signal steady and analyse it properly. This is very useful in automotive diagnostics for correctly reading signals from various sensors and devices.

The trigger synchronises the oscilloscope's **sweep** with the signal being checked. By setting a trigger point, the oscilloscope shows a steady waveform matched to a specific event, such as the signal's rising or falling edge. This matching makes viewing clear and steady, helping to measure and analyse the waveform accurately.

How to set a trigger

Setup

Step 1
- Select Trigger Type: Choose the appropriate trigger type based on the signal to be analysed. Common types include edge, level, and pulse width.

Step 2
- Adjust Trigger Level: Set the trigger level to the specific voltage at which the oscilloscope will activate. This is typically done by adjusting a control button, drag and drop or entering a value on the touchscreen interface.

Step 3
- Select Trigger Slope: Determine whether you want the oscilloscope to trigger on a rising edge (positive slope) or a falling edge (negative slope) of the signal.

Step 4
- Choose Trigger Mode: Select the trigger mode that best suits the measurement needs. Options often include auto, normal, single, and rapid modes.

Stabilising waveforms using edge or level triggers

Edge triggers are the most commonly used type in automotive diagnostics. They trigger the oscilloscope based on the selected signal's rising or falling edges. By setting the trigger level and slope, you can stabilise periodic waveforms, making it easier to identify patterns and anomalies. (It appears to hold the waveform stationary). Level triggers are similar but focus on the signal achieving a specific voltage level, regardless of the slope. This is especially useful for signals that do not have prominent edges but maintain consistent voltage levels. **Table 3.1** describes various trigger modes.

Setup and Use

Table 3.1 Trigger modes

Trigger	Description
Auto mode	In auto mode, the oscilloscope continuously sweeps and updates the display even if no trigger event occurs. This mode ensures that the screen is always refreshed, useful for initial signal checks.
Normal mode	In normal mode, the oscilloscope only sweeps and updates the display when a trigger event occurs. This mode is useful for capturing infrequent or single-shot events, ensuring that only relevant data is displayed.
Single mode	Single mode captures a single waveform and then stops. This mode is ideal for capturing **transient events** or anomalies that do not repeat, allowing for detailed analysis without the waveform scrolling off the screen. It is also useful for capturing individual diagnostic tasks where the start of the event is triggered by a rise or fall in voltage, such as relative compression.
Rapid mode	Rapid mode captures multiple waveforms in quick succession, providing a series of frames for analysis. This mode is useful for examining variations in signals over a short period.

Using a single trigger to capture a discrete diagnostic measurement that occurs when switching a circuit on or cranking an engine for example, allows you to focus on the act of creating the conditions for the test, rather than worrying about scope settings.
It works by moving the refresh point from the far side of the screen further to the right. Normally when the trace reaches the righthand side of the screen, it refreshes and starts again at the lefthand side. Because the trace cannot effectively reach its refresh point, it produces a condition where the capture runs continuously but looks as though the trace has paused. As soon as the voltage changes when the circuit is switched, the trace begins to move across the screen until it reaches the trigger point, where it stops showing one screen of capture.

Trigger - a signal or event that initiates a specific action within a vehicle's electronic or diagnostic system.

Sweep - the continuous movement of an oscilloscope's display trace across the screen, capturing and visualising electrical signals over time.

Transient events - short-lived changes in voltage, current, or signals within a vehicle's electrical system.

Think of an automotive oscilloscope trigger like a camera shutter:
- The oscilloscope waits for the exact moment a specific event happens—just like you wait for the right lighting or motion before snapping a picture.
- Once triggered, it captures the waveform at the precise point, ensuring a stable and repeatable view (snapshot) of the signal.

Without a trigger, the waveform would be a blurry, unpredictable image—just like snapping a photo without a steady hand.

Setup and Use

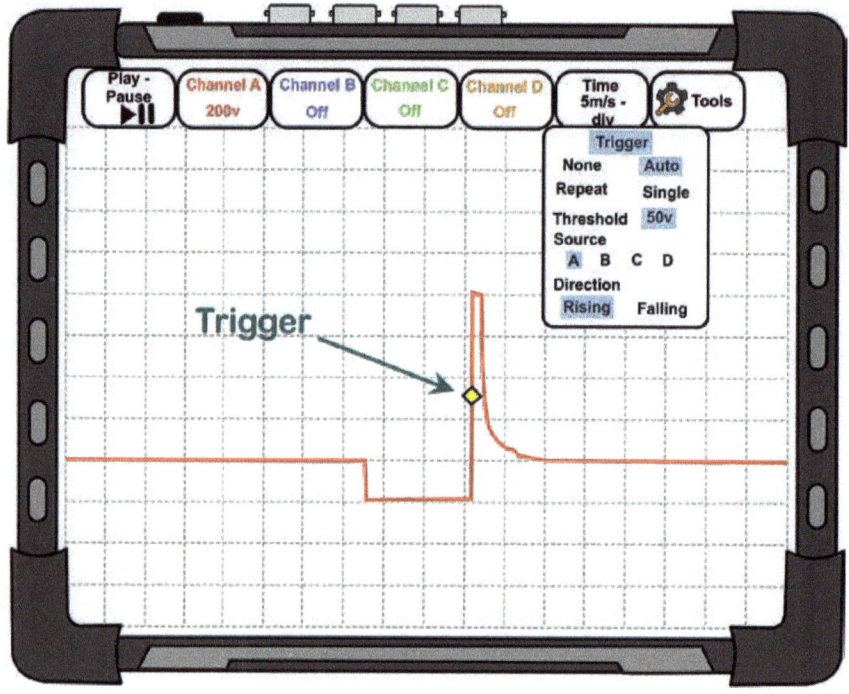

Figure 3.9 Triggers

Table 3.2 provides some examples of using triggers when diagnosing automotive electrical faults.

Table 3.2 Using triggers when diagnosing electrical faults

Use	Diagnosis
Diagnosing a faulty crankshaft sensor	By using an edge trigger on the rising edge of the crankshaft sensor signal, you can stabilise the waveform and observe any inconsistencies or missing pulses that might indicate a faulty sensor.
Analysing ignition coil waveforms	Setting a level trigger at the expected voltage level of the ignition coil's primary waveform allows you to observe the coil's behaviour, ensuring proper operation and identifying issues like misfiring or weak spark generation.
Inspecting fuel injector signals	Using a rapid trigger mode to capture successive injector pulses helps when comparing the performance of multiple injectors. This is crucial for identifying an injector that may be delivering inconsistent fuel quantities. (This might need current and voltage pintle humps to be compared).
Monitoring alternator output	An auto trigger mode can provide a continuous view of the alternator output waveform, allowing you to monitor charging system performance and identify issues like voltage drops or irregular output.

Practical Usage

Acquiring waveforms

Acquiring accurate waveforms is essential for diagnosing issues, ensuring proper operation, and optimising performance. This section provides a guide to capturing signals from sensors, actuators, and circuits using automotive oscilloscopes.

Setup and Use

Understanding automotive signals

Before delving into waveform acquisition, it is crucial to understand the types of signals encountered in automotive systems.

These signals can be divided into three main categories:

- Sensors: Devices that monitor conditions such as temperature, pressure, position, and speed. Examples include crankshaft position sensors, camshaft position sensors, mass airflow (MAF) sensors, and oxygen (O2) sensors.
- Actuators: Components that perform actions based on signals received from the engine control unit (ECU). Examples include fuel injectors, ignition coils, solenoids, and relays.
- Circuits: Electrical pathways that connect sensors and actuators to the ECU and other control units. These circuits include power supply lines, ground connections, and signal transmission lines.

An oscilloscope is a great way of quickly accessing and checking circuit and component operation. They can often be a good option for **first-look** diagnostics. A simple connection to the suspected circuit can be used to see if the system has a 'pulse', allowing you to potentially rule in or rule out components and focus your systematic diagnostic routine.

Setup and preparation

To acquire accurate waveforms, proper setup and preparation are essential.

Setup

Step 1
- Choose an oscilloscope that matches the requirements of your diagnostics. Consider the bandwidth, sample rate, number of channels, and triggering capabilities. Automotive oscilloscopes designed for vehicle diagnostics will often have specialised features for capturing and analysing automotive signals.

Step 2
- Use appropriate probes and leads for your oscilloscope. Ensure that they are in good condition and free of damage. Connect the probe to the component or circuit you wish to test, following any available manufacturer guidelines for placement and connection. Common probe types include voltage probes, current probes, and specialty probes for specific sensors. Ensure your connection techniques do not cause damage to any wiring, cables, connectors or terminals.

Step 3
- Configure the oscilloscope's trigger mode to stabilise the waveform and capture the signals accurately. Adjust the trigger level to match the expected signal amplitude to prevent capturing noise or **extraneous signals**.

Step 4
- Proper grounding is crucial to avoid signal noise and ensure accurate waveform capture. Connect the oscilloscope ground lead to a clean ground point on the vehicle. Avoid grounding to noisy or high-current points.

Step 5
- Once the oscilloscope is set up, operate the circuit or system and obtain a waveform by adjusting the amplitude and time base. Set triggers to stabilise the waveform and pause the capture to compare with known good examples. Add cursors or rulers to take measurements and record your results for future reference.

Setup and Use

While the battery's negative terminal is generally adequate for **differential voltage** measurements, it can sometimes introduce noise or interference.
Whenever feasible, connect to the sensor or actuator ground instead, as this often provides a more stable and cleaner waveform.

Practical usage tips

First look diagnosis

First look diagnosis is a quick way to check the overall health of vehicle systems without taking it apart. By using an automotive oscilloscope, you can measure things like compression, ignition, and fuel system signals through simple tests—often just by connecting to the battery with an amp clamp or using a pressure sensor. The oscilloscope shows live waveforms on a screen, helping spot problems like misfires, weak compression, or timing issues early, saving time and guiding more detailed checks.

First-look diagnosis can serve as an efficient method to quickly determine whether vehicle-related issues are due to mechanical or electrical faults. This approach can streamline your diagnostic process and enhance overall efficiency.

For instance:
Consider an electric window that is malfunctioning. The issue could stem from either an electrical problem or a physical mechanical fault.
Place an inductive amp clamp around one of the battery wires (the choice of wire is irrelevant) and setup the oscilloscope to measure current with a time base of 1 s/div.
Activate the faulty electric window and observe the trace.
- If no amplitude change is detected, the fault is likely an electrical open circuit, allowing you to focus your diagnostic efforts on this issue.
- If the trace exhibits a high amplitude (current draw), it suggests that the window motor is experiencing excessive strain, potentially due to a worn motor, seized regulator, or runners. Your diagnostics can now concentrate on this mechanical problem.
- If the trace shows a low amplitude (current draw) compared to other windows, it indicates electrical high resistance. Now your diagnostic procedures can be directed towards resolving this electrical issue.

This method can be adapted to various systems under test, enabling differentiation between electrical and physical faults without the necessity for extensive disassembly.

Acquiring clean signals

Ensure clean signal capture by avoiding interference from neighbouring components and circuits. Use **shielded probes** and leads to minimise noise.

Interpreting waveforms

Analyse captured waveforms to identify patterns and abnormalities. Compare waveforms against known good patterns and manufacturer specifications to diagnose issues accurately. The companion book, *Automotive Oscilloscopes Waveform Analysis*, provides examples of common automotive waveforms and explains their attributes.

Setup and Use

Saving and exporting data

Export the data to a computer or storage device for further study and documentation. Understanding the types of signals, setting up the equipment properly, and following best practices will enable you to capture accurate waveforms and diagnose automotive issues efficiently.

Why save oscilloscope data?

- Future Reference – Comparing current waveforms to past captures allows you to spot patterns, trends, and gradual system degradation.
- Verification and Documentation – Saved waveforms serve as objective records, supporting repair decisions and customer communication.
- Training and Knowledge Sharing – A well-maintained waveform database can be used to educate other technicians, improving overall diagnostic capabilities.
- Troubleshooting Recurring Issues – If an intermittent fault reoccurs, referencing previous captures can accelerate diagnosis, and reduce guesswork.

By implementing organised saving strategies, you create a living diagnostic tool that helps solve problems faster and with greater accuracy, ensuring high-quality vehicle repairs and ongoing professional development.

Where possible, it is good practice to compare waveforms with known good examples to help identify anomalies; this is how the waveform should look.
Although looking at faulty examples can be useful, they can also be misleading. The innumerable number of faulty waveforms possible may lead to the chance of misinterpreting the data and discounting a potential issue.
Therefore, '**known goods**' are best.

First look - a rapid, initial evaluation of a vehicle's health using non-intrusive diagnostic tools, aimed at identifying potential issues by analysing electrical signals, system performance, and mechanical symptoms without disassembling components.

Extraneous signals - unwanted electrical noise or interference captured by an oscilloscope that can obscure or distort the accurate representation of a vehicle's system signals during diagnostics.

Differential voltage - the voltage difference measured between two points in a vehicle's electrical circuit, used to diagnose system performance or detect faults.

Shielded probes - a specialised oscilloscope probe designed to minimise electromagnetic interference (EMI) and noise when measuring signals in vehicles.

Known goods - reference signals from properly functioning components or systems, used as benchmarks during oscilloscope diagnostics to compare and identify faults in a vehicle's electrical or mechanical performance.

Setup and Use

Common Automotive Applications

Ignition waveforms

For petrol-operated engines, the ignition system is a critical component responsible for igniting the air-fuel mixture in the engine's cylinders to produce power. Accurate analysis of ignition waveforms can provide valuable insights into the performance and health of this crucial system. When using an oscilloscope to diagnose ignition-related issues, understanding the distinct waveforms associated with spark plugs, coils, and older distributors is essential.

Spark Plug Signal Analysis: The waveform of spark plugs typically shows a high-voltage spike, followed by a rapid drop and subsequent oscillations. This high-voltage spike, known as the firing line, occurs when the ignition coil discharges, creating a spark across the spark plug gap. The subsequent oscillations, or **ringing**, result from the ignition coil's secondary winding reacting to the discharge. By analysing the height, width, and shape of the firing line, you can assess the ignition system's efficiency, detect misfires, and identify issues such as worn spark plugs, fouling, or incorrect gaps.

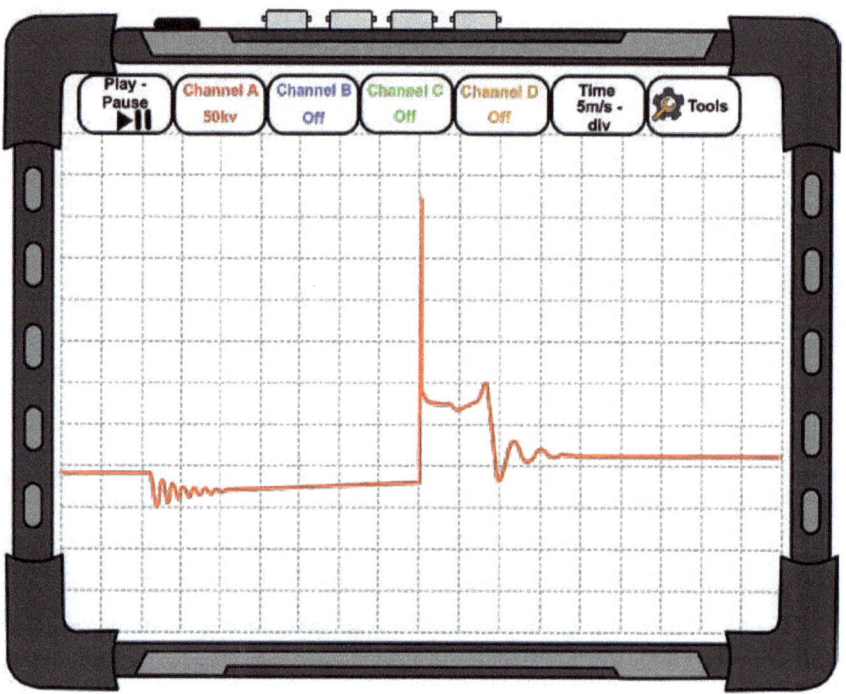

Figure 3.10 Secondary ignition waveform

Coil Signal Analysis: Modern vehicles often use coil-on-plug (COP) or coil-near-plug (CNP) systems, eliminating the need for a distributor. Each coil's primary and secondary waveforms can be analysed to evaluate their performance. The primary waveform, seen on the low-voltage side of the coil, shows the voltage supplied to the coil and the control module's switching events. The secondary waveform, on the high-voltage side, reflects the energy transferred from the coil to the spark plug. Variations in these waveforms can indicate issues such as coil degradation, short circuits, or problems with the control module.

Setup and Use

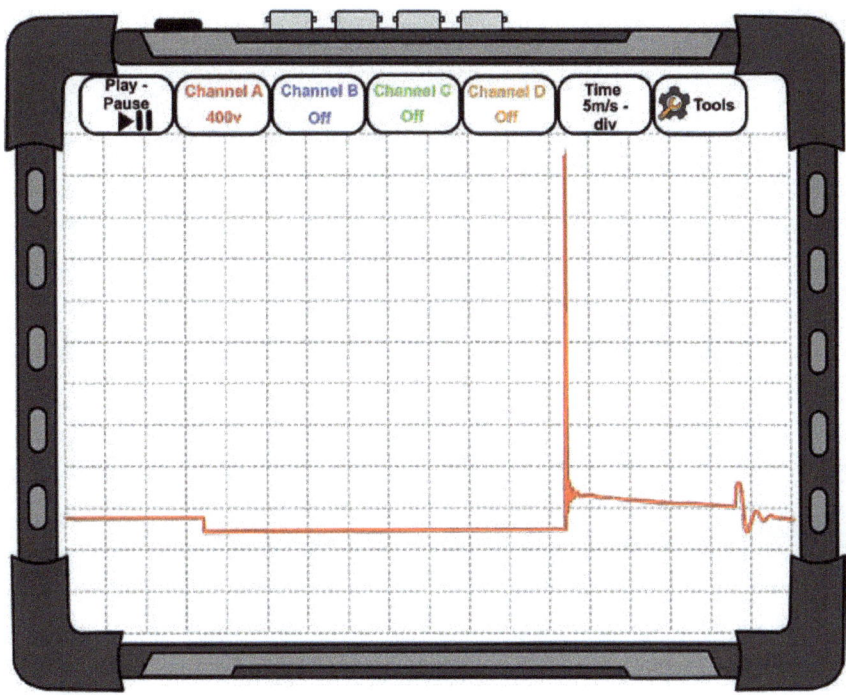

Figure 3.11 Primary ignition waveform

Distributor Signal Analysis: In older ignition systems with distributors, waveform analysis focuses on the distributor's role in directing high-voltage pulses to the correct spark plug. The distributor waveform typically shows a series of high-voltage peaks corresponding to each cylinder's firing. By examining the timing and consistency of these peaks, you can diagnose distributor-related issues, including worn distributor caps, rotor problems, or misalignment.

Sensor diagnostics

Testing crankshaft, camshaft, MAF, and O2 sensors

In modern automotive systems, sensors play a crucial role in monitoring various parameters to ensure optimal performance, efficiency, and safety. Diagnostics using an oscilloscope enable accurate analysis of sensor signals and identification of issues that may affect vehicle functionality. Here are a few examples of key sensors, including crankshaft, camshaft, mass airflow (MAF), and oxygen (O2) sensors.

Crankshaft position sensor

The crankshaft position sensor monitors the position and rotational speed of the crankshaft, providing critical data for engine timing and fuel injection. During diagnostics, the sensor signal can be captured and analysed to ensure it accurately reflects the crankshaft's motion. The waveform should display consistent pulses corresponding to each crankshaft revolution. Irregularities in the signal may indicate sensor malfunctions, wiring issues, a misfire, or problems with the crankshaft itself. Compare the readings to the manufacturer's specifications or known good waveform examples for accurate diagnosis.

Setup and Use

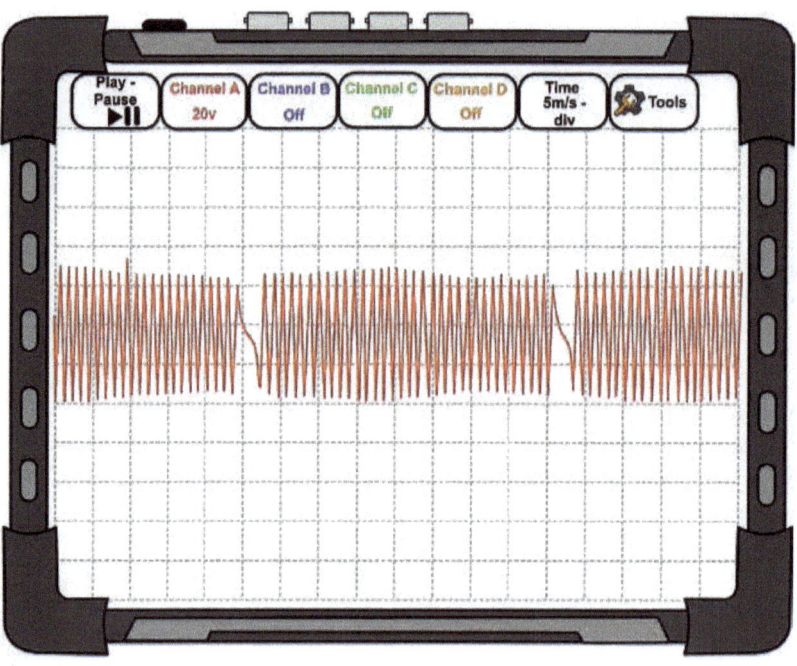

Figure 3.12 Crankshaft position sensor waveform

Camshaft position sensor

The camshaft position sensor works alongside the crankshaft sensor to synchronise fuel injection and ignition timing. It is also used by some engine management systems to assist with variable valve timing, indicating position relative to the combustion cycle. Analysing the camshaft sensor signal involves checking for uniformity and alignment with the crankshaft signal. Correct correlation will require a known good capture as a reference. Discrepancies can indicate timing chain or belt issues, sensor faults, or camshaft wear. Compare the readings to the manufacturer's specifications or known good waveform examples for accurate diagnostics.

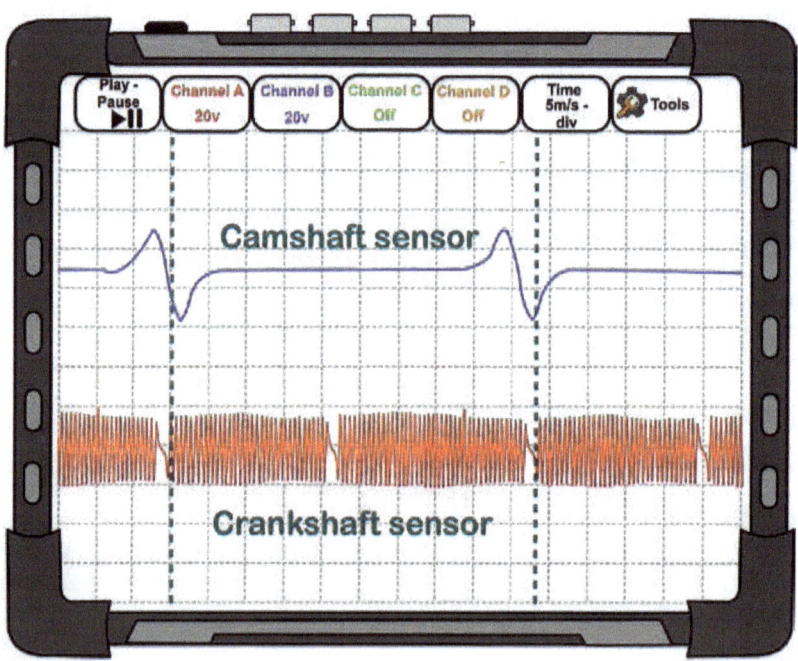

Figure 3.13 Camshaft position sensor waveform

Setup and Use

Mass air flow (MAF) sensor

The MAF sensor measures the amount of air entering the engine, which is vital for proper fuel mixture and combustion. During diagnostics, the oscilloscope can capture the sensor output, which should show a proportional response to changes in airflow. Mechanical intake pulses will often create minor fluctuations in the general signal, however, overall, the waveform should have a relatively smooth rise and fall related to throttle opening and closing. Abnormalities such as spikes, drops, or excessive noise may indicate contamination, sensor degradation, or air leaks in the intake system for example. Compare the readings to the manufacturer's specifications or known good waveform examples for accurate diagnostics.

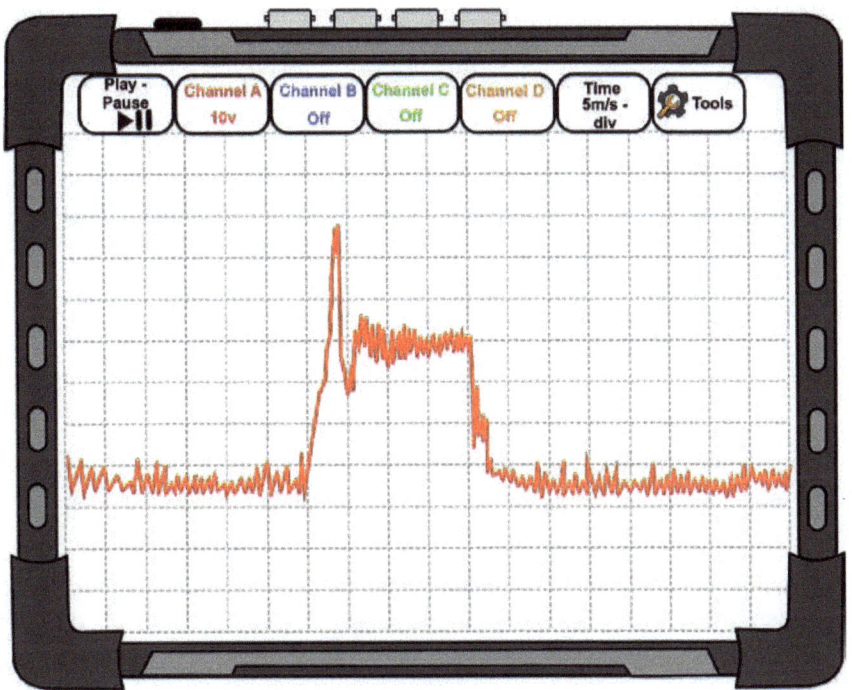

Figure 3.14 Mass air flow sensor waveform

Oxygen (O2) sensor

Oxygen sensors monitor the oxygen levels in the exhaust gases, providing the engine control unit (ECU) with data to adjust the air-fuel ratio. Using an oscilloscope, the sensor signal should show oscillations between lean and rich conditions, indicating efficient combustion. A slow response, flat lines, or inconsistent oscillations can indicate sensor ageing, exhaust leaks, or fuel system problems. Compare the readings to the manufacturer's specifications or known good waveform examples for accurate diagnostics.

Setup and Use

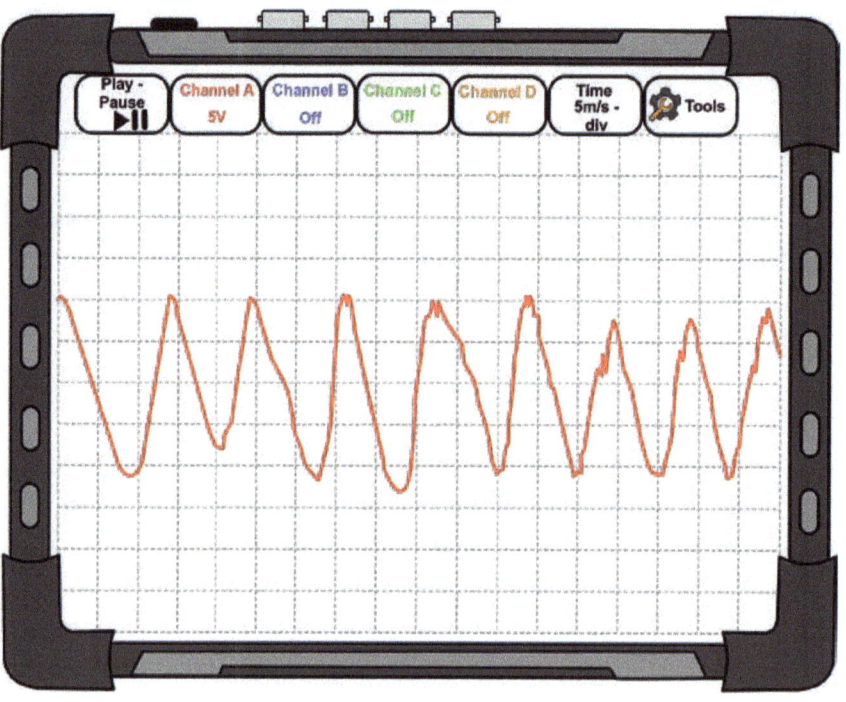

Figure 3.15 Oxygen sensor waveform

 Pre and post catalytic converter oxygen sensor waveforms should be different from one another.
Due to effective operation of the catalyst, the oxygen content will potentially be less following conversion, because of the chemical process. This means that the post catalyst oxygen sensor waveform received will have a lower amplitude (height), than the pre-catalyst one. This provides the engine management system with an indication of catalyst efficiency.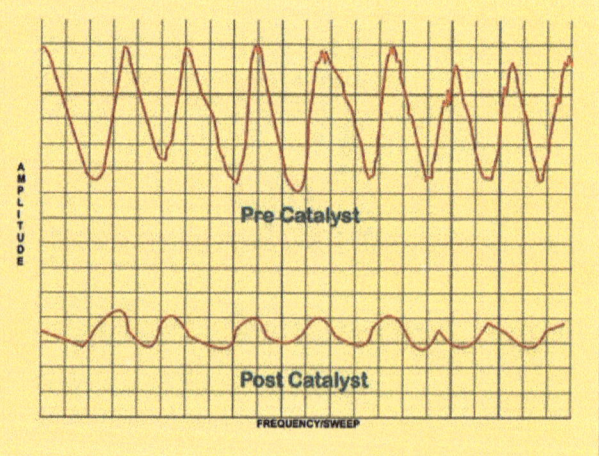

 Broadband or wideband exhaust oxygen sensors produce a relatively flat wave or curve relative to the air-fuel ratio because of their design and function. To observe their operation using an oscilloscope, it may be necessary to artificially change the air-fuel ratio by altering the fuel pressure or introducing an intake air leak. Care must be taken when conducting this form of diagnosis as it could potentially create running issues or even cause damage. Do not conduct this style of testing if there is a risk of personal injury or potential damage to the vehicle or equipment. After any testing, always scan for and clear any diagnostic trouble codes (DTC) that may have been created by the diagnostic routine.

Setup and Use

Modern vehicles are equipped with an array of sensors to monitor and optimise performance. Although not exhaustive, here are some examples of potential sensors that you may encounter:

Petrol Vehicles

- Mass Air Flow (MAF) Sensor
- Oxygen Sensor
- Throttle Position Sensor
- Knock Sensor
- Coolant Temperature Sensor
- Crankshaft Position Sensor
- Camshaft Position Sensor
- Manifold Absolute Pressure (MAP) Sensor
- Fuel Pressure Sensor
- Oil Pressure Sensor
- Ambient Air Temperature Sensor
- Vehicle Speed Sensor

Diesel Vehicles

- Diesel Particulate Filter (DPF) Sensor
- Exhaust Gas Recirculation (EGR) Sensor
- Boost Pressure Sensor
- Glow Plug Temperature Sensor
- Fuel Rail Pressure Sensor
- NOx Sensor
- Air Intake Temperature Sensor
- Differential Pressure Sensor
- Turbo Speed/Wastegate Sensor

Electric Vehicles

- Battery Management System (BMS) Sensors
- Motor Position Sensor
- Inverter Temperature Sensor
- Charging Port Sensor
- Regenerative Braking Sensor
- High Voltage Current Sensor
- Cabin Climate Sensors
- Wheel Speed Sensors
- Proximity Sensors
- Thermal Management Sensors

ADAS

- Camera Sensors
- Radar Sensors
- LiDAR Sensors
- Ultrasonic Sensors
- GPS/GNSS Sensors
- Inertial Measurement Units (IMU)
- Driver Monitoring Sensors
- Wheel speed sensors
- Steering angle sensors
- Brake pressure sensors
- Accelerometers
- Decelerometers
- Crash sensors

Actuators

Actuators play a crucial role in modern automotive systems, converting electrical signals from control units into physical actions. Diagnosing these components with an oscilloscope enables you to precisely analyse actuator signals and identify issues affecting vehicle functionality. Here are examples of common actuators, including fuel injectors, EGR valves, and throttle bodies.

Fuel injectors

Fuel injectors are responsible for delivering the correct amount of fuel into the engine cylinders at precise intervals. Proper functioning of fuel injectors is crucial for optimal engine performance and fuel efficiency. The style and type of waveform will vary depending on design and use, such as petrol port injection, petrol direct injection, high-pressure diesel common rail injection, and Pumpe-Düse (PD).
A typical injector waveform includes a pulse indicating the opening and closing of the injector valve. Analyse the pulse width and frequency to determine injector performance. Deviations may indicate issues such as clogging, electrical faults, or mechanical wear. Compare the readings to the manufacturer's specifications or known good waveform examples for accurate diagnostics.

Setup and Use

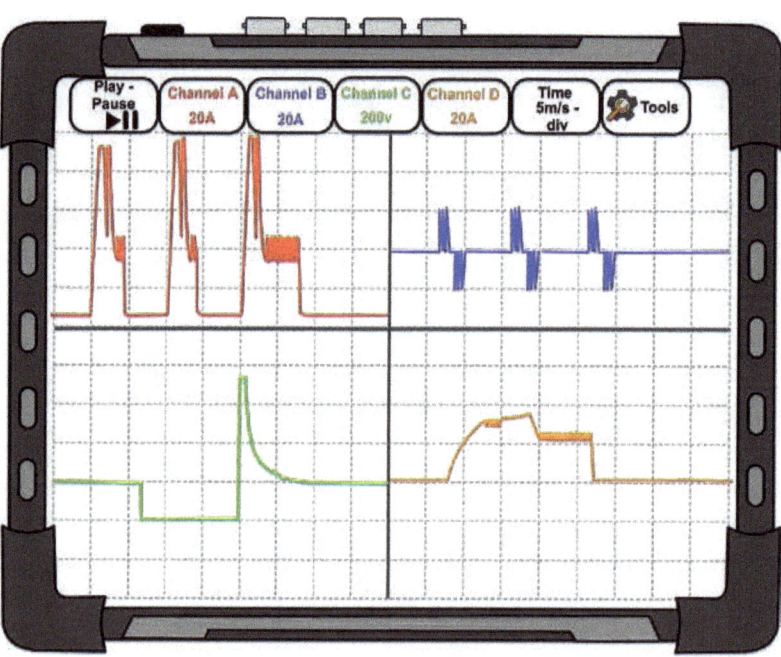

Figure 3.16 Fuel injector waveforms

EGR valves

Exhaust Gas Recirculation (EGR) valves assist in reducing nitrogen oxide emissions by recirculating a portion of the exhaust gas back into the engine intake. Proper EGR function is crucial for meeting emission regulations and maintaining engine efficiency. The waveform should display consistent pulses (typically square waves) corresponding to the valve's opening and closing cycles. These pulses will indicate a duty cycle, which should vary with engine operation and be proportional to the EGR valve position. Irregularities in the waveform may signal issues such as valve sticking, electrical problems, or clogging. Compare the readings to the manufacturer's specifications or known good waveform examples for accurate diagnostics.

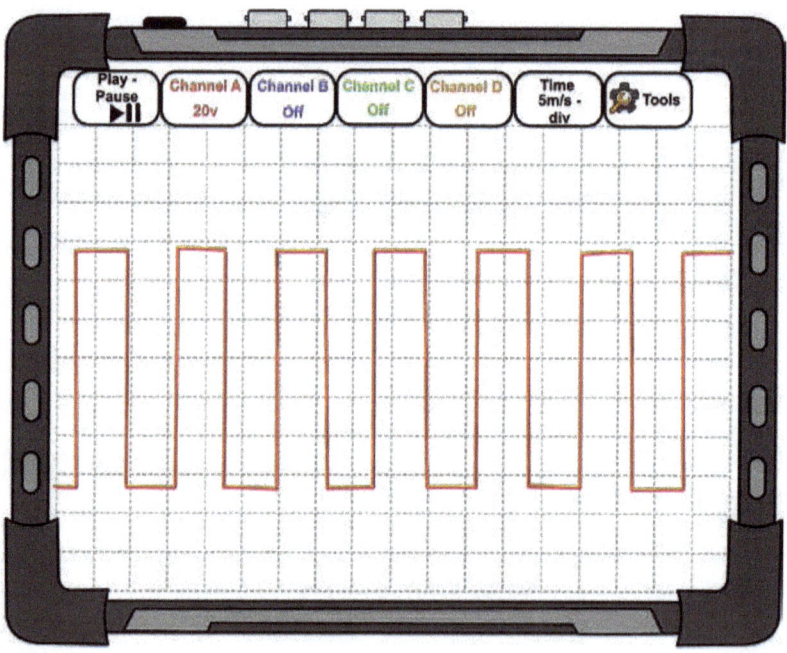

Figure 3.17 EGR waveform

Setup and Use

Throttle bodies

Throttle bodies regulate the amount of air entering the engine based on driver input. Accurate throttle body operation is essential for smooth engine performance and acceleration. Power up the engine and observe the waveform. The waveform reflects the throttle position, conveying real-time data on its movements. Abnormal patterns in the waveform may suggest issues such as sensor malfunction, wiring problems, or mechanical failures. Compare the readings to the manufacturer's specifications or known good waveform examples for accurate diagnosis.

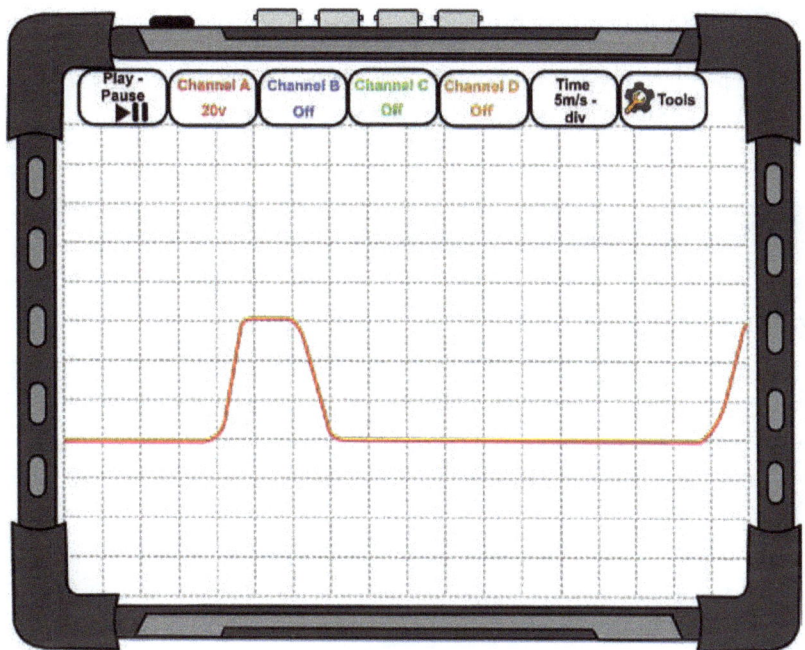

Figure 3.18 Throttle body waveform

Modern vehicles are equipped with an array of actuators to provide functions, maintain and operate components. Although not exhaustive, here are some examples of potential actuators that you may encounter:

- Fuel Injectors
- EGR Valves
- Throttle Bodies
- Idle Air Control Valves
- Variable Valve Timing (VVT) Solenoids
- Transmission Solenoids
- ABS Actuators
- HVAC Blend Door Actuators
- Turbocharger Wastegate Actuators
- Electric Water Pumps
- Cooling Fan Actuators
- Electronic Parking Brakes
- Window and Door Lock Actuators
- Headlight Levelling Actuators
- Fuel Pump Actuators

Voltage and current testing

Understanding voltage and current testing is crucial for anyone who aims to master the use of automotive oscilloscopes. This section looks at the specifics of measuring the operation of starter motors, alternator output, and detecting parasitic drain.

Setup and Use

Measuring starter motor operation

To accurately measure the starter motor operation, it's essential to understand the relationship between voltage drop and current draw during engine cranking. An oscilloscope can be used to capture the voltage drop across the starter motor.
Connect the oscilloscope voltage probes to the battery terminals and observe the drop as the engine is cranked. Add a current probe to the main starter motor cable to observe the current draw while cranking; this will show the initial current required to overcome **inertia** and current versus the compression loads.
A healthy starter motor will show a consistent pattern, while irregularities may indicate issues such as worn brushes, a failing armature, or even low cylinder compression. Compare the readings to the manufacturer's specifications or known good waveform examples for accurate diagnosis.

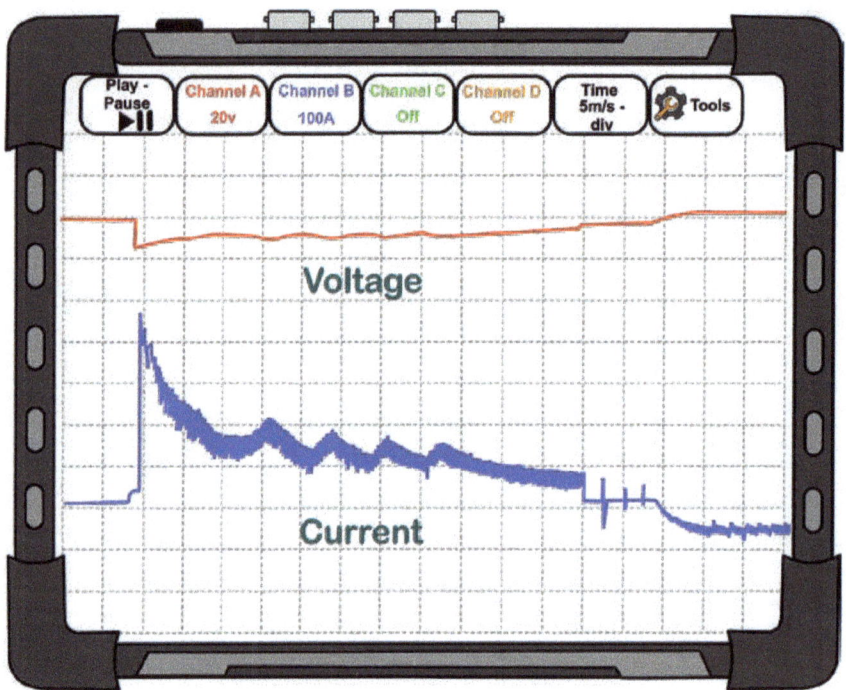

Figure 3.19 Starter voltage versus current

 Relative compression testing is a valuable diagnostic technique, helping you to assess the health of an engine's internal components without the need for intrusive procedures. By using an oscilloscope and an amp probe, you can quickly determine whether each cylinder is contributing equally to the engine's operation.

A relative compression test measures the current draw of the starter motor while cranking the engine. The current drawn by the starter motor increases when it encounters resistance, such as during the compression stroke of each cylinder. By analysing the waveform generated by the oscilloscope, you can identify cylinders with low compression, which may indicate issues such as worn piston rings, damaged valves, or head gasket failure.

Setup and Use

Setup and use

Step 1
- Preparation: Ensure the battery is fully charged and the engine is at **ambient temperature**. Disable the fuel or ignition system to prevent the engine from starting during the test.

Step 2
- Connecting the Amp Probe: Attach an amp probe to the oscilloscope and clamp it around the positive battery cable or the main feed wire to the starter motor. Ensure the amp probe is securely connected to capture accurate readings.

Step 3
- Initial Cranking: Instruct a helper to crank the engine while you observe the waveform on the oscilloscope or use a single trigger to create a capture.

Step 4
- Capture: The oscilloscope should display a series of peaks corresponding to the current draw of each cylinder during its compression stroke.

Step 5
- Analysing the Waveform: Compare the height of each peak. Ideally, the peaks should be of similar height, indicating even compression across all cylinders. Variations in peak height suggest compression issues in the corresponding cylinders.

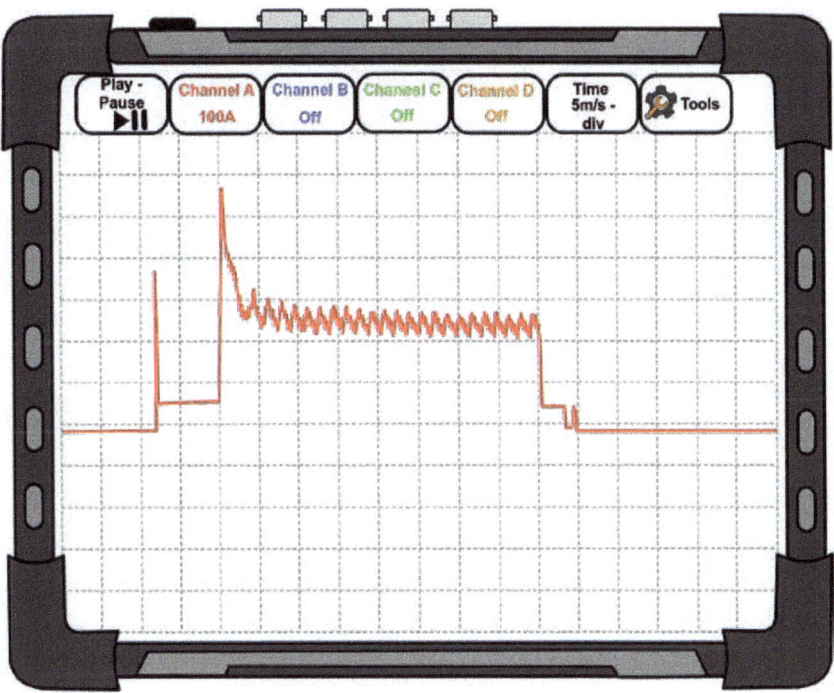

Figure 3.20 Relative compression

Setup and Use

An amp clamp/probe often has arrows moulded into the clamp or body to indicate the direction of current flow. These arrows should be positioned towards or away from the battery to produce either a positive or negative waveform. If the waveform capture appears inverted, turning the probe 180 degrees will correct this issue.

If the waveform shows consistent peaks of equal height, it provides a quick indication that all cylinders have similar compression levels, suggesting healthy engine operation. However, if one or more peaks are significantly lower than the others, it points to possible compression issues in those cylinders. Further diagnostics, such as a cylinder leak-down test or in-cylinder pressure analysis may be needed to pinpoint the exact cause of the low compression.

A second channel can be connected to the fuel injection, ignition or camshaft sensor and synchronised with the relative compression waveform to help identify which cylinder has the low compression without any further dismantling.

Measuring alternator output

The alternator is vital for maintaining the vehicle's electrical system and battery charge. To measure alternator output using an oscilloscope:

- Connect the voltage probes to the battery terminals.
- Place a current probe around one of the battery cables.

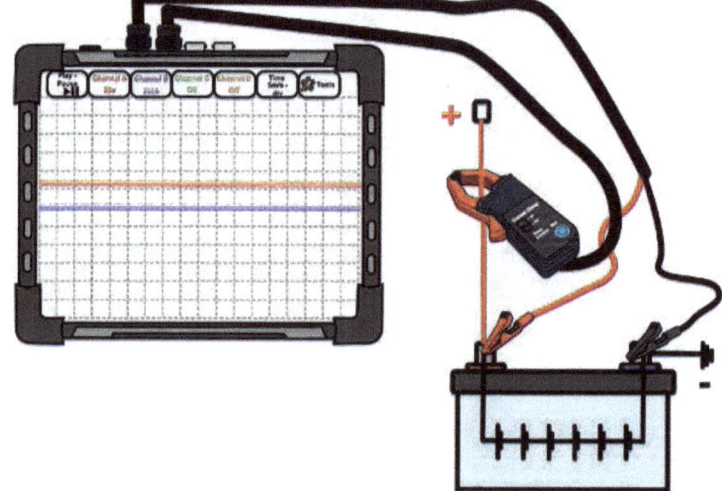

Figure 3.21 Measuring alternator output

You should typically see a stable output voltage that is higher than battery potential, with minor fluctuations. You should also see a fairly smooth current output, which indicates that the **rectifier** is functioning correctly.
If there are significant deviations, this could suggest issues such as a faulty voltage regulator or problems with the alternator diodes. Compare the readings to the manufacturer's specifications or known good waveform examples for accurate diagnosis.

Setup and Use

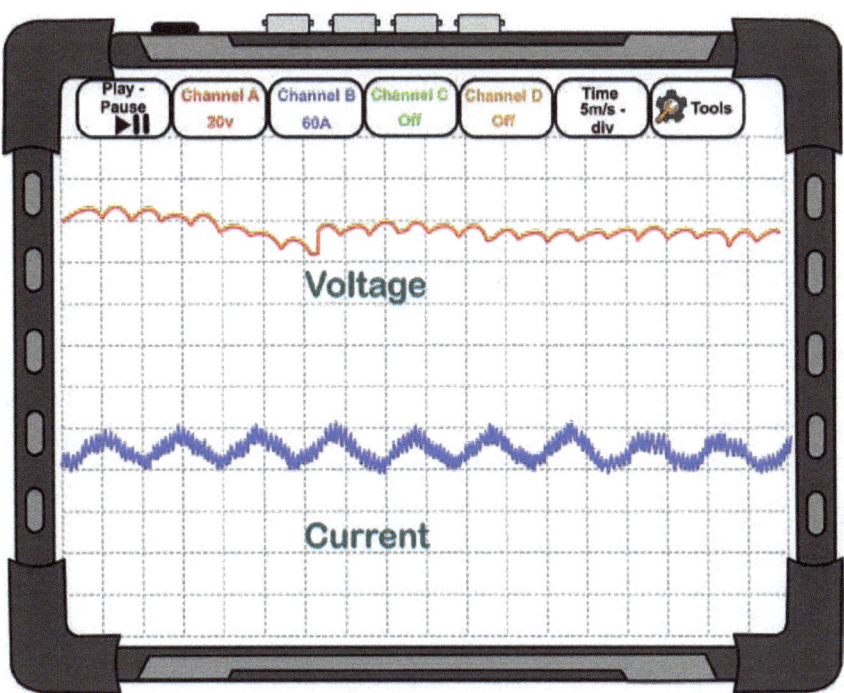

Figure 3.22 Alternator output

Detecting parasitic draws

Parasitic draws can drain a vehicle's battery when the engine is off, which can lead to starting problems. To detect parasitic draws, set your oscilloscope to measure current and place a current clamp around the battery's positive or negative cable. Ensure that all circuits are turned off where possible and allow any in-vehicle networks to go to sleep. Observe the waveform for any continuous current draw, which should be minimal when the vehicle is off. To monitor current drain over a longer period of time, set a very slow timescale, allowing identification of intermittent parasitic drains. Identifying and tracing the source of the draw is essential to effectively resolve battery drain issues.

Inertia - a bodies resistance to changes in motion due to its mass.

Ambient temperature - the external temperature surrounding a vehicle.

Rectifier - a component that converts AC (alternating current) from the alternator into DC (direct current) to charge the battery and power electrical systems.

Parasitic draw - the unwanted continuous drain on a vehicle's battery when the engine is off.

In-vehicle networks

Controller Area Network (CAN) BUS systems are integral to automotive electrical systems, allowing communication between various electronic control units (ECUs) in vehicles. Understanding and diagnosing CAN BUS-related issues require precise tools, such as the oscilloscope, to capture and analyse data signals.

Setup and Use

Setup and use

Step 1: Connect the oscilloscope probes to the CAN BUS High (CAN_H) and CAN BUS Low (CAN_L) lines.

Step 2: Observe the waveform patterns, which should appear as opposing differential voltage signals oscillating between CAN_H and CAN_L.

Step 3: Look for any distortions or irregularities in the signal that may indicate faults such as short circuits, open circuits, or interference.

Step 4: Use serial decoding functions available in advanced oscilloscopes to translate the CAN BUS signal into readable data frames and analyse for error frames, overload frames, or any anomalies indicating network issues.

Step 5: Compare the readings to the manufacturer's specifications or known good waveform examples for accurate diagnostics.

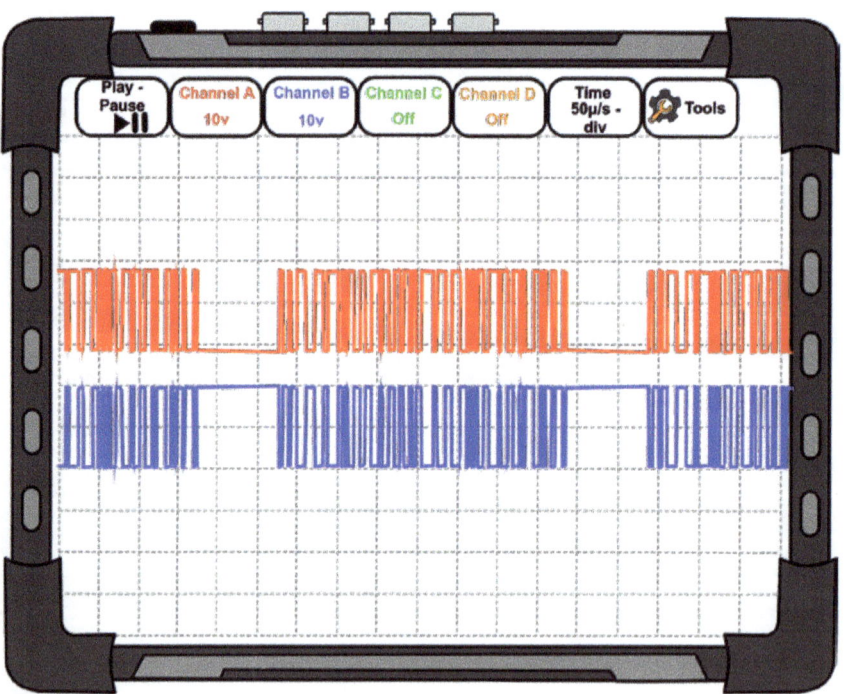

Figure 3.23 CAN BUS waveform

Setup and Use

Using an oscilloscope to measure vehicle mechanical systems

In-cylinder compression testing with a pressure transducer

An oscilloscope, paired with a pressure transducer, offers a precise method for evaluating in-cylinder compression, which is crucial for diagnosing engine performance issues. Connect the pressure transducer to the oscilloscope, ensuring compatibility between the two devices to obtain accurate pressure readings. Start or run the engine and observe the waveform on the oscilloscope. The waveform should display peaks and troughs corresponding to the intake, compression, and exhaust strokes. Valve timing and other mechanical issues can also be analysed. Compare the readings to the manufacturer's specifications or to known good waveform examples for accurate diagnosis.

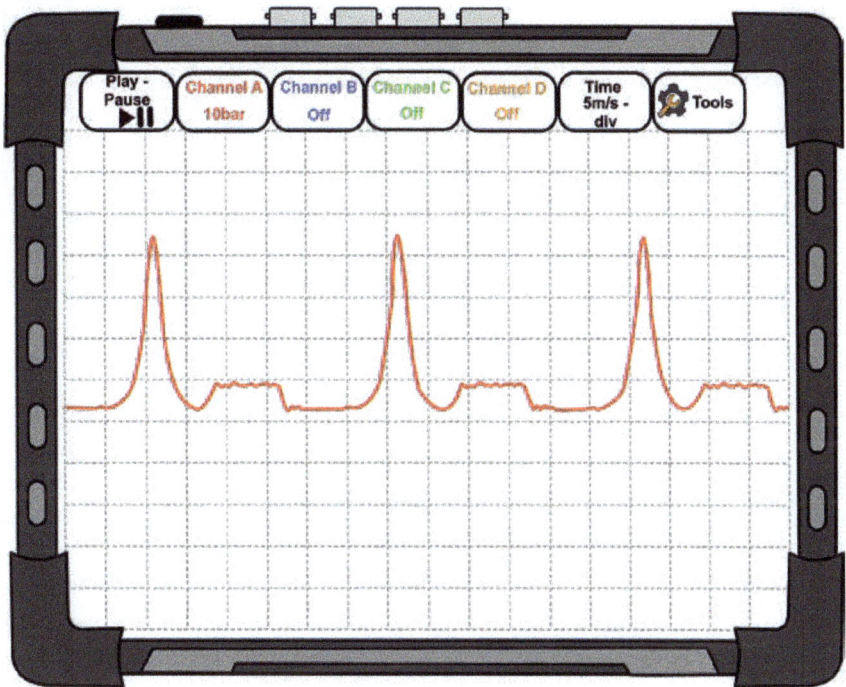

Figure 3.24 In-cylinder pressure waveform

Intake and exhaust pulse analysis

Intake and exhaust pulse analysis is another critical diagnostic tool that can be performed using an oscilloscope and a pressure transducer.

- Intake Pulse Analysis: Connect the pressure transducer to the intake manifold. Observe the intake pulses on the oscilloscope. These pulses provide insight into valve condition/operation and the integrity of the intake system.
- Exhaust Pulse Analysis: Similarly, connect the pressure transducer to the exhaust tailpipe. The exhaust pulses reveal information about the exhaust valve operation and can help identify issues such as restricted exhaust flow or misfire problems.

Compare the readings to the manufacturer's specifications or known good waveform examples for accurate diagnosis.

Setup and Use

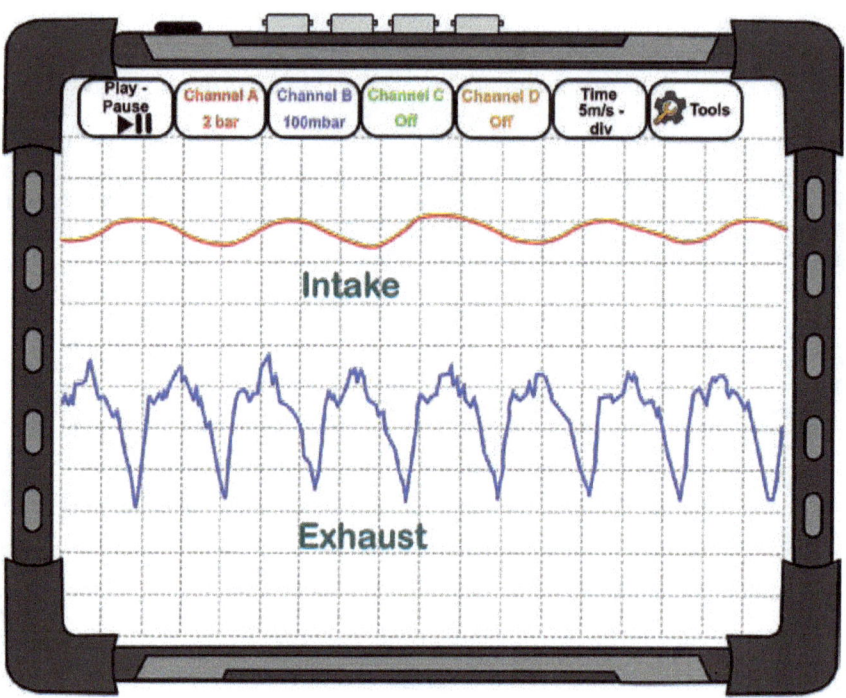

Figure 3.25 Intake and exhaust pressure waveforms

NVH diagnostics for chassis systems

Noise, Vibration, and Harshness (NVH) are indicators of potential issues within the engine, transmission, chassis or body system. By using an oscilloscope to measure NVH, you can diagnose faults more effectively. This will require the use of specific NVH probes and diagnostic software.

- Vibration Analysis: Attach accelerometers to various components of the vehicle's engine, transmission or chassis. Connect these sensors to the oscilloscope to capture vibration data. Analyse the frequency and amplitude of the vibrations to pinpoint the source of issues, such as unbalanced and worn engine, transmission and suspension components.
- Noise Analysis: Use microphones placed in strategic locations to capture noise data. The oscilloscope can display these signals, allowing for the identification of abnormal noise patterns that may indicate problems like engine, body or drivetrain issues.
- Harshness Analysis: Measure the overall vehicle response to road inputs by capturing data from sensors placed on the suspension and chassis. Analyse the data to identify areas where the ride quality is compromised, indicating potential faults in the shock absorbers, springs, or other components.

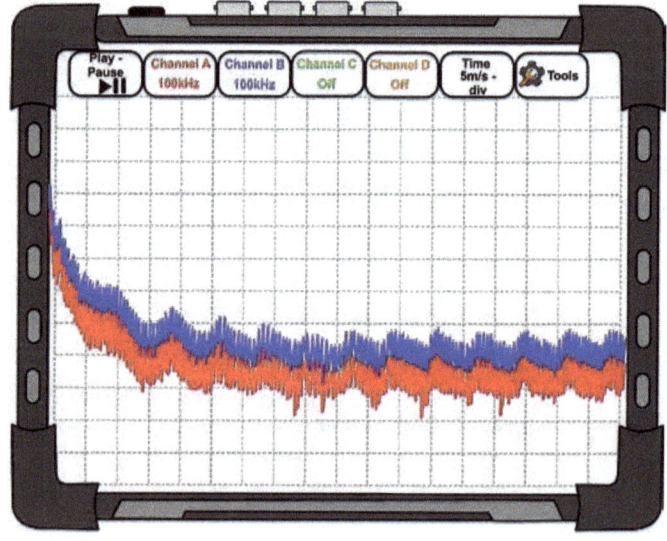

Figure 3.26 NVH waveform

Setup and Use

Troubleshooting and Maintenance

Avoiding signal noise

To ensure the accuracy of diagnostics using oscilloscopes, it is crucial to avoid signal noise during the capture process. Signal noise can distort readings and lead to incorrect conclusions about the condition of vehicle components. **Table 3.3** shows examples of best practices for achieving clean signal capture.

Table 3.3 Best practice for signal capture

Use high-quality probes and cables	Invest in high-quality probes and cables designed specifically for automotive oscilloscopes. Ensure that all connectors are secure and free from wear or damage. High-quality accessories reduce the likelihood of interference and provide more accurate signal readings.
Proper grounding	Ensure that the oscilloscope and the vehicle are properly grounded. For straightforward measurements involving differential voltage, connect the oscilloscope's ground probe to a clean, unpainted metal surface of the vehicle or the battery negative post. Poor grounding can introduce noise and lead to unreliable data. Where possible, the signal ground from the component or circuit being tested often provides the cleanest waveform.
Minimise electromagnetic interference (EMI)	Position the oscilloscope and probes away from sources of **electromagnetic interference EMI** or **radio frequency interference RFI**, such as ignition systems, alternators, and other high-voltage components. Use shielded cables to further reduce the impact of EMI on signal quality.
Signal conditioning	Use signal conditioning techniques, such as **filtering**, to minimise the impact of noise. Digital filters can be applied to remove unwanted frequencies from the signal, enhancing clarity and accuracy.
Maintain a stable environment	Conduct tests in a stable environment to minimise external factors that might affect the readings. Avoid performing diagnostics in areas with heavy machinery or electrical equipment that can introduce external noise and interference into the system.
Regular calibration	Regularly check the calibration of your oscilloscope and probes to ensure they are functioning correctly and providing accurate readings. Calibration helps maintain the integrity of the signal and the reliability of the diagnostic results. If the probe relies on an internal battery (a current clamp for example), ensure that it is in good condition, and replace immediately if a low battery warning is displayed.
Proper probe placement	Place probes precisely as per the diagnostic requirements. Incorrect probe placement can introduce signal noise and lead to misinterpretation of the data. Follow manufacturer guidelines for optimal probe positioning.

Electromagnetic Interference EMI - the disruption of electrical systems caused by electromagnetic signals or noise.

Radio Frequency Interference RFI - the disturbance in vehicle electronics caused by unintended radio frequency signals.

Filtering - the process of refining signal data by removing unwanted noise or artifacts from waveforms.

Setup and Use

Recognising faulty probes

Proper probe function is essential for the accurate operation of oscilloscopes. Faulty probes can lead to unreliable data and misdiagnosis, which may result in improper repairs and maintenance. To ensure the reliability of diagnostic results, it is crucial to identify and replace damaged probes promptly.

Signs of a faulty probe

There are several indicators that a probe may be faulty:

- Inconsistent Readings: If the oscilloscope provides erratic or fluctuating readings that do not correspond with expected signal patterns, the probe may be damaged.
- Physical Damage: Inspect the probe for visible signs of wear and tear, such as frayed wires, cracked insulation, or bent connectors. Physical damage can compromise the probe's integrity and signal transmission.
- Signal Noise: A damaged probe may introduce excessive noise into the signal, resulting in unclear or distorted waveforms. Compare readings with a known-good probe to verify.
- Intermittent Connectivity: If the signal intermittently drops or cuts out, it may be due to a faulty connection within the probe.

Oscilloscope maintenance

Proper maintenance and cleaning of your oscilloscope and its probes are essential for ensuring long-term performance and diagnostic accuracy. As a precision instrument, the oscilloscope must be handled with care to avoid damage and maintain its functionality. **Table 3.4** provides some key practices for maintaining your oscilloscope.

Table 3.4 Maintaining your oscilloscope

Regular cleaning	Dust and debris can accumulate on the oscilloscope's exterior and connectors, leading to poor connections and potential malfunctions. Use a soft, dry cloth to wipe down the exterior surfaces of the oscilloscope regularly. For more thorough cleaning, a slightly dampened cloth with mild soap and water can be used but ensure that no moisture enters the device.
Connector care	The connectors on your oscilloscope and probes are critical for accurate signal transmission. Inspect them regularly for signs of dirt, corrosion, or damage. Clean the connectors using a suitable electronic contact cleaner and a lint-free cloth. Avoid using abrasive materials that could damage the contact surfaces.
Probe maintenance	Probes are delicate components that require careful handling. Avoid pulling, twisting, or bending the probe cables excessively, as this can lead to internal wire breakage. When not in use, store the probes in a protective case to shield them from dust, moisture, and physical impacts. Periodically inspect the probes for signs of wear, such as frayed cables or damaged connectors, and replace them if necessary.
Environmental considerations	Store and use the oscilloscope in a clean, dry, and stable environment. Extreme temperatures, humidity, and vibrations can adversely affect the performance of the oscilloscope. If the oscilloscope is used in harsh environments, consider using protective covers and enclosures to shield it from environmental stress.
Software Updates	Keep the oscilloscope's firmware and software up to date by regularly checking the manufacturer's website for updates. Software updates can provide improved functionality, bug fixes, and enhanced performance, ensuring that your oscilloscope remains at peak operational efficiency.

Setup and Use

 Mixing probes or cables from different manufacturers can cause inconsistent results. For example, when testing a CAN BUS system, if two different test leads are connected to CAN High and CAN Low, internal resistances in the cables could cause discrepancies with signal speed, edge matching or even amplitude.

Safety Considerations

Voltage limits

Respecting voltage limits when using an oscilloscope is crucial for equipment safety and measurement accuracy. Automotive systems, especially in hybrid and electric vehicles, often involve high voltages. Exceeding your oscilloscope's voltage rating can damage the device, cause costly repairs, or endanger the operator.
To avoid damage, always check the maximum voltage rating of your oscilloscope and probes before connecting them to a circuit. This information is typically provided in the user manual or on the manufacturer's website. Ensure that the voltage of the signals you intend to measure does not exceed these ratings. If you are unsure about the voltage levels in a particular circuit, use a multimeter to measure them before connecting your oscilloscope.
Additionally, consider using attenuating probes or differential probes designed to handle higher voltages. These specialised probes reduce the signal amplitude before it reaches the oscilloscope, allowing you to safely measure higher-voltage signals without exceeding the oscilloscope's input limits. Attenuators are particularly useful in automotive applications where voltage spikes can occur.
Always follow proper grounding procedures to prevent accidental short circuits or electric shocks. Ensure that the oscilloscope and the vehicle are properly grounded to avoid creating dangerous potential differences that could damage the equipment or harm the user.

Preventing short circuits and probe placement

When using an oscilloscope, the correct placement of probes is crucial to prevent short circuits, ensure accurate measurements, and reduce the possibility of damage to the vehicle or components. Improper probe placement can lead to unintended connections between different parts of the circuit, potentially damaging the oscilloscope, the vehicle's electrical system, or even causing injury to the operator.
Here are key considerations for safe probe placement:

- Identify Test Points: Before connecting any probes, identify the specific test points within the circuit. Ensure that these points are accessible and that connecting probes will not interfere with other components or wiring.
- Use Insulated Probes: Where possible, always use probes that are well-insulated to prevent accidental contact with adjacent components. Insulated probes reduce the risk of creating unintended short circuits, which can be particularly dangerous when dealing with high-voltage automotive systems.
- Secure Probes Properly: Secure the probes firmly to the test points using suitable connectors or clips. Loose connections can lead to intermittent contact, resulting in erratic readings and the potential for sparks or shorts.
- Avoid Crossing Wires: When routing probe cables, avoid crossing them over or under other wires and components. This precaution helps prevent accidental disconnections and minimises electrical interference that could distort the oscilloscope readings.
- Use Breakouts: If available, use breakout boxes or leads which allow non-intrusive connection to an electrical circuit or component.
- Check for Voltage Ratings: Verify that the probes and oscilloscope are rated (category or CAT) for the voltage levels present in the circuit. Using probes with insufficient voltage ratings can result in potentially dangerous insulation breaches.

Setup and Use

Safe handling of high voltages

High voltage systems are common in modern vehicles, from ignition systems that generate sparks to electric vehicles that operate on high-voltage batteries. These systems pose significant risks if not handled properly, including electrical shock, damage to diagnostic equipment, and potential harm to the vehicle's electronic components.

Ignition systems

Automotive ignition systems produce voltages in tens of thousands of volts, which are necessary for creating the spark that ignites the fuel-air mixture in the engine. These high voltages can be dangerous, leading to electrical shocks or damage to equipment if precautions are not taken.

Electric vehicles

Electric vehicles (EVs) operate on high-voltage batteries, often ranging from 200 to 800 volts. These voltages are sufficient to cause fatal electric shocks if mishandled. Additionally, EVs have complex electrical architectures that require careful probe placement and insulation to avoid accidental shorts and ensure safety.

When working with hybrid and electric vehicles, it's crucial to be aware of the dangers of electricity, including the risk of electric shock or electrocution. Since electricity is invisible, it can only be detected with specialised equipment. Here are some key electrical units to remember:

- **Voltage**: This is the electrical pressure or force.
- **Amperage**: Also known as electric current, this is the quantity of flowing electricity.
- **Resistance**: This is a restriction in electric current flow, measured in ohms.
- **Power**: This is the rate at which work is done, measured in watts.

All these units influence the risk of electric shock or electrocution. However, voltage is the primary factor to consider.

The human body naturally resists electrical voltage up to a certain limit. When this limit is exceeded, the voltage induces a potentially harmful current flow. It's important to note that while voltage can be dangerous, it's the current that causes damage.

The level of voltage considered dangerous varies among individuals due to differences in personal electrical resistance and points of contact with an electrical circuit. Therefore, various values may be quoted when researched, but caution is advised when dealing with any voltage.

For dry, unbroken human skin, the touch threshold is often cited as 50 volts DC or 25 volts AC. This value can decrease if the skin is wet or if the points of electrical contact penetrate the skin. Once the voltage touch threshold is reached, a current as small as 80 milliamps (a milliamp is one-thousandth of an amp) can be fatal under certain circumstances. According to vehicle regulations, ECE R-100 paragraph 2.14 defines high-voltage as a classification for an electrical component or circuit. It's considered high-voltage if its working voltage exceeds 60 volts DC but is less than 1500 volts DC, or if it's over 30 volts AC but less than 1000 volts AC RMS.

Setup and Use

Other dangers include:

Short circuit
Electrocution can happen at stated voltage levels due to the body's internal resistance. However, a short circuit, created by bridging high and low voltage with a metal conductor, can discharge electric current rapidly, causing heat, sparks, fire, or explosions, even below safe touch voltage. Therefore, any electrical voltage is hazardous. Use insulated tools around these systems to reduce short circuit risks.

Magnetic fields
Electric vehicles generate strong magnetic fields around high-voltage cables, which can pose risks to people with electronic life-sustaining devices like pacemakers and insulin pumps. While it's generally safe to drive these vehicles, working on their high-voltage systems is not recommended for individuals with such health conditions.

Personal Protective Equipment (PPE)

Always wear appropriate PPE when working with high-voltage systems. This includes insulated gloves, protective eyewear, and clothing that covers exposed skin. PPE provides a barrier against accidental contact with high-voltage components. **Table 3.5** provides some examples of personal protective equipment recommended when working on the systems of high voltage electric vehicles.

Table 3.5 Examples of high-voltage personal protective equipment PPE

PPE	Recommendations and use
High-voltage insulated gloves and over gauntlets 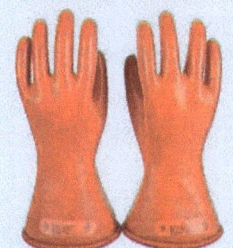 	High-voltage PPE gloves are designed to protect the hands from electric shock when working with high-voltage systems. High-voltage PPE gloves are usually made of rubber or synthetic materials that have high dielectric strength, which means they can resist the flow of electric current. • High-voltage PPE gloves are supplied in different class categories based on the maximum voltage they are rated to withstand, ranging from Class 00 (500 volts) to Class 4 (36,000 volts). The minimum class required for working on and around the high-voltage systems of electric vehicles is Class 0 (1,000 volts). • High-voltage PPE gloves should be worn with leather protectors over them to prevent damage from abrasion, cuts, or punctures. However, they provide no extra protection against electric shock, so they should be used in addition to, but not instead of, high-voltage insulated gloves. • High-voltage PPE gloves need to be tested before each use to ensure they are free of defects or holes that could compromise their insulation. The gloves can be inflated at the wrist and then rolled up to ensure that no air leaks are present. • When a new pair of gloves is commissioned, the date they are brought into use should be recorded. Once exposed to air and UV, the materials begin to degrade. Many glove manufacturers will stipulate that the gloves will require recertification or replacement after 6 months.

Setup and Use

Table 3.5 Examples of high-voltage personal protective equipment PPE

PPE	Recommendations and use
Eye protection 	Eye protection is available in various formats. It is primarily designed to provide protection against impact, heat, chemicals, and fumes. The most appropriate form of eye protection should be chosen for the type of task being undertaken. • Safety glasses are a type of protective eyewear designed to protect your eyes from small flying objects, such as splinters or dust, as well as chemicals and light. They usually have side shields or wrap around the temples to prevent objects from entering from the sides. • Safety goggles are a type of protective eyewear designed to protect your eyes from small flying objects, such as splinters or dust, as well as chemicals and light. They fit tightly against the face and form a seal around the eyes, preventing any particles or liquids from entering. • Safety face shields are a type of protective eyewear designed to protect your face from small flying objects, chemicals, light, and heat. They can be worn over glasses or goggles to provide extra protection.
Overalls or workwear 	Specific workwear or overalls provide an additional layer of protection between the user and potential hazards. Depending on the materials used, this protection could include containment of loose clothing and the covering of areas of bare skin, resistance to chemicals or limited fire protection. It is recommended that overalls or clothing do not use metal fastenings when working on or around the high-voltage systems of hybrid and electric vehicles. Metal fastenings may increase the risk of short circuit, leading to electrocution or fire, as well as interfere with the very strong magnetic fields found within the drive systems of electric vehicles. Arc flash/blast overalls are a type of PPE that is designed to protect the wearer from the hazards of arc flash and blast, which are intense bursts of heat, light and sound that can occur when an electrical fault causes an arc between two conductors. Arc flash and blast can cause severe burns, blindness, hearing loss, shock and even death. Arc flash/blast overalls are usually made of flame-resistant fabrics that do not ignite, melt, or drip when exposed to high temperatures. They may also have reflective strips or patches to increase visibility in low-light conditions.

Setup and Use

Table 3.5 Examples of high-voltage personal protective equipment PPE

PPE	Recommendations and use
Safety footwear and high-voltage overshoes  	Safety footwear is a type of PPE that is designed to protect your feet from injuries caused by impact, penetration, heat, cold, chemicals or electricity. Safety footwear is usually made of leather, rubber or synthetic materials that have different properties and resistance levels against various hazards. Some common features of safety footwear include: • Protective toe caps: These are made of steel, aluminium, composite, or plastic materials that can withstand high forces and prevent crushing or puncturing of the toe area. • Penetration-resistant midsoles: These are made of steel, textile or composite materials that can prevent sharp objects from piercing through the sole of the shoe. • Electrically insulated soles: These are made of non-conductive materials that can inhibit electricity and reduce the possibility of electric shock. They are also often marked with a yellow or green triangle symbol to indicate their electrical safety rating. High-voltage overshoes are a type of protective footwear designed to protect feet from electric shock when working with high-voltage systems. High-voltage overshoes are usually made of rubber or synthetic materials that have high dielectric strength, which means they can resist the flow of electric current. High-voltage overshoes are worn over regular shoes and cover the entire foot area. Similar to high-voltage insulated gloves, the overshoes have different class categories based on the maximum voltage they can withstand, ranging from Class 00 (500 volts) to Class 4 (36,000 volts). The minimum class required for working on and around the high-voltage systems of electric vehicles is Class 0 (1,000 volts).

Voltage rating verification

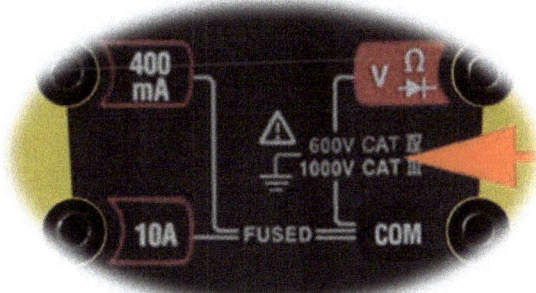

Verify that the probes and the oscilloscope are rated for the voltage levels present in the circuit. Using equipment with insufficient voltage ratings can lead to insulation breakdown and dangerous situations. The minimum recommended rating for electrical test equipment used on the high voltage systems of electric vehicles is CATIII 1000 volts.

Figure 3.27 Correctly rated tooling

Setup and Use

Following any repairs or replacement of high-voltage electrical components, diagnoses and tests, you need to conduct checks to ensure the serviceability, safety and operation of the electrical systems.

Indirect tests using vehicle information systems and/or live data should be perfectly safe for you to conduct. However, you must always view the reliability and accuracy of any data obtained on its merits and the source of information provided.

Only people who have had sufficient training and experience should conduct physical measurements using diagnostic test tools, especially if 'live' high-voltage readings or checks are required.

Diagnostic equipment and electrical test measurement tools must be correctly rated for the voltages to be tested, calibrated, and checked against a known good voltage source for accuracy.

You must always wear high-voltage personal protective equipment (PPE).

Handling high voltages in automotive diagnostics requires a meticulous approach to safety. By adhering to these protocols, you can ensure accurate measurements and prevent accidents while working with the sophisticated electrical systems of modern vehicles. Safety is paramount, and with proper precautions, high-voltage diagnostics can be performed effectively and securely.

Conclusion

The configuration of automotive oscilloscopes is crucial to their functionality and effectiveness. It is advisable to frequently practice using your oscilloscope, and if time permits, experiment with various settings while it is connected to an active circuit. This approach will help you gain a deeper understanding of the settings and their impact on waveform captures, thereby enhancing your expertise and confidence in operating the device.

Oscilloscope Technology and Features

Chapter 4 Oscilloscope Technology and Features

Automotive oscilloscopes have revolutionised the field of vehicle diagnostics, offering unparalleled insights into the intricate workings of modern automotive electrical systems. These sophisticated instruments are no longer reserved for elite engineers but have become indispensable tools for mechanics and technicians striving to maintain the highest standards of vehicle performance and safety. This chapter will describe automotive oscilloscope technology and features. It is essential to understand both the foundational principles and advanced capabilities these devices offer.

Contents

Advanced Oscilloscope Architecture	**94**
Enhanced Triggering Techniques	**96**
High-Speed Sampling	**100**
Memory Depth Management	**102**
Specialised Setup Techniques	**103**
Integration with Other Tools	**111**

The automotive industry is a high-risk environment, especially when dealing with electrical systems. The hazards of electricity are well-known but can be easily ignored due to its invisible nature. This can lead to complacency if the fundamentals of electricity are not well understood. Even with this understanding, caution is necessary. Assume that any safety systems designed for protection have failed and take precautions to minimise the risk of injury or death. Always evaluate the risks associated with any activity and implement measures to eliminate or reduce the hazards involved in any task, diagnosis, or repair. Additional risks associated with working on, or around electrical systems may include:

- Electrocution
- Strong magnetic fields
- Falling from heights
- Short circuits
- Electrical discharge/arcing
- Fire and explosion
- Chemicals

Oscilloscope Technology and Features

Advanced Oscilloscope Architecture

Modern **digital** oscilloscopes are indispensable tools when used in automotive diagnostics. They allow you to capture and analyse complex electrical signals with remarkable precision. At the heart of these sophisticated devices lies an intricate **architecture** designed to convert, process, and display electrical waveforms for detailed examination.

The journey of a signal through a digital oscilloscope begins with the input stage, where the signal is received via probes. These probes must accurately capture the voltage levels without significant distortion or loss. Once the signal is acquired, it passes through an **attenuator** and **amplifier**, which adjust the signal's **amplitude** to match the oscilloscope's input range. This ensures that the signal is neither too weak to detect nor too strong to cause distortion.

The next component in the architecture is the analogue-to-digital converter (ADC). The ADC samples the **analogue** signal at a high rate, transforming it into a series of digital values that the oscilloscope can process.

The sampling rate is a key performance metric. A higher sampling rate allows the oscilloscope to capture fast-changing signals with greater accuracy, preserving the integrity of high-frequency components.

Once **digitised**, the signal enters the processing stage, where it is stored in the oscilloscope's memory. The depth of this memory determines how much data can be captured in a single sweep. Greater memory depth allows for longer signal captures, which can be crucial when diagnosing intermittent faults or analysing lengthy **communication protocols**.

Within the processing unit, the oscilloscope uses various **algorithms** to interpret the digital data. This includes the application of **filters** to remove noise, enhancing the clarity of the signal. The oscilloscope can perform mathematical operations on the data, such as **Fast Fourier Transform** (FFT), to transition from time-domain analysis to frequency-domain analysis. This capability is essential for diagnosing issues related to signal integrity and electromagnetic interference (EMI).

The triggering system within the oscilloscope is another vital aspect of its architecture. Advanced triggering techniques enable the oscilloscope to precisely capture events of interest. **Triggers** can be set based on standard parameters such as voltage thresholds, or more complex conditions like pulse width, **runt signals**, and specific patterns. This ensures that the oscilloscope only records relevant data, making the analysis more efficient.

The final stage in the signal's journey is the display system. Modern oscilloscopes feature high-resolution screens or are connected to computers and laptops that provide a clear and detailed view of the waveforms. The user interface is designed to be intuitive, offering various tools for **zooming**, measuring, and **annotating** the signal data. This visual representation is invaluable, allowing you to pinpoint **anomalies** and diagnose issues effectively.

Think of an oscilloscope like your body's reflex system.
- When you touch something hot, sensory neurons send a signal to your brain—just like a probe picks up an electrical signal.
- Your brain processes the information, deciding that immediate action is needed—this is the oscilloscope's internal circuitry analysing the waveform.
- Then, your brain sends a signal to your muscles, making you pull your hand away—just as the oscilloscope displays the processed waveform, showing you exactly what happened in the circuit.

Fast, precise, and essential for diagnosing what's happening in the moment.

Oscilloscope Technology and Features

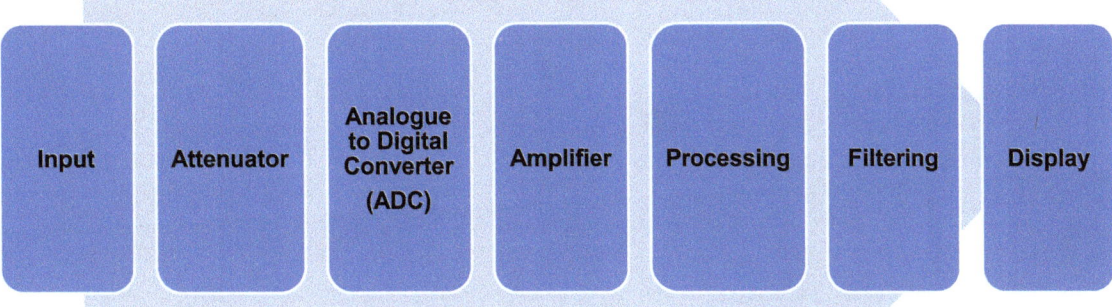

Figure 4.1 The signal journey

Digital - information using binary code (0s and 1s) for processing, storage, or transmission in electronic systems.

Architecture - the integrated framework of electronic components, wiring, and systems within an oscilloscope that enables power distribution, signal transmission, and communication between various electrical and digital systems for functionality and control.

Attenuator - a device that reduces or regulates the strength of signals, such as sound, electrical currents, or radio frequencies.

Amplifier - an electronic device that increases the strength or magnitude of an input signal, such as sound, voltage, or current.

Amplitude - the magnitude or strength of a signal, such as voltage or sound wave, within a vehicle's electronic or communication systems.

Analogue - a continuous signal or representation that varies in magnitude or frequency to convey information.

Digitised - the conversion of analogue information into digital format using binary code (0s and 1s) for electronic processing, storage, or transmission.

Communication Protocols - standardised rules and formats that enable reliable data exchange between electronic control units (ECUs) and systems within a vehicle, such as CAN, LIN, or FlexRay, ensuring seamless communication and functionality.

Oscilloscope Technology and Features

Algorithms - step-by-step instructions designed to process data and perform tasks in vehicle systems, such as controlling engine performance, enabling driver assistance features, or optimising energy efficiency.

Filters - tools within an oscilloscope that refine or smooth signal waveforms by reducing noise or unwanted frequencies, enabling clearer analysis of a vehicle's electrical signals.

Fast Fourier Transform (FFT) - a mathematical algorithm used to analyse and convert time-domain signals into their frequency-domain components, enabling diagnostics and signal processing in vehicle electronic systems.

Triggers - settings or mechanisms that stabilise and define when an oscilloscope captures and displays a signal, ensuring precise analysis of specific events in a vehicle's electrical systems.

Runt signals - low-amplitude signals in a vehicle's electronic systems that fail to reach the expected voltage or strength, often indicating irregularities or faults in the circuitry.

Zooming - the capability of an oscilloscope to magnify specific sections of a signal waveform for detailed analysis, aiding in the precise examination of electrical signals in vehicle systems.

Annotating - the act of adding explanatory notes, comments, or highlights to a text, image, or document to provide clarification, insights, or context.

Anomalies - deviations or irregularities from the expected norm, pattern, or standard, often indicating errors, unusual events, or abnormal conditions in data, systems, or behaviour.

Enhanced Triggering Techniques

Understanding automotive oscilloscope triggers

Oscilloscope triggers are settings or mechanisms designed to stabilise the display of signals and determine when signals are captured. They ensure that particular events in a vehicle's electrical systems are captured at just the right time and accurately analysed, helping you to diagnose and troubleshoot effectively.

Think of an oscilloscope trigger like a starting gun in a race.

- If the gun fires too soon, the racers might not be ready—just like an oscilloscope triggering on random noise before a meaningful signal appears.
- If the gun fires too late, the key moment—the runners launching forward—gets missed (false start).

Setting the trigger correctly ensures that the oscilloscope captures the exact moment the critical signal occurs, just as a well-timed starting gun ensures the race begins at the right instant.

Oscilloscope Technology and Features

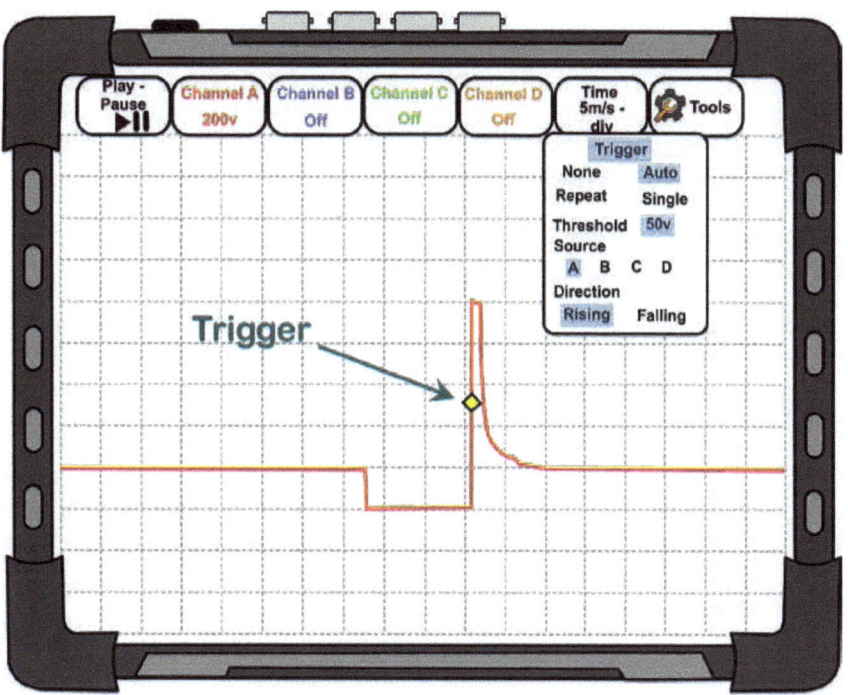

Figure 4.2 Trigger

Trigger types

Auto trigger

The Auto Trigger mode is designed to continuously capture and display waveforms, irrespective of whether a specific trigger event has occurred. This mode ensures that the oscilloscope screen is never blank, providing a constant view of the signal.

Auto Trigger is particularly useful for observing signals that are relatively stable or repetitive, where continuous monitoring is required.

Advantages	Disadvantages
• It allows for uninterrupted signal display, aiding in the identification of trends or variations over time. • Auto trigger eliminates the need for precise manual adjustments, making it easier for you to quickly observe waveforms without fine-tuning the trigger settings. • When analysing signals with long intervals between events (such as sensor outputs at idle), auto trigger ensures a stable display without waiting for a specific trigger condition.	• As it does not wait for a trigger event, this mode may display irrelevant or noisy data, making it less suitable for capturing sporadic or specific events. • Auto trigger can introduce variations in where the waveform starts on the screen, making it harder to precisely time events like injector pulses or ignition sequences.

Repeat trigger

Repeat Trigger mode enables the oscilloscope to continuously capture, and display signals based on recurring trigger events. Unlike the Auto Trigger, the Repeat Trigger requires a specific condition to be met before capturing the signal, ensuring more relevant data is displayed.

This mode is beneficial for monitoring regular but intermittent signals, such as those in cyclic automotive systems like ignition or fuel injection.

Advantages	Disadvantages
• It provides a balance between continuous data display and relevance, capturing signals only when the trigger condition is satisfied. • Unlike Auto Trigger, Repeat Trigger locks onto a valid signal event each time, reducing jitter and improving waveform clarity for detailed analysis.	• It may miss transient or very rare events that occur outside the regular intervals of the trigger condition. • To get the best results, you must set trigger levels and conditions correctly, which can take time and might be challenging for beginners.

Single trigger

Single Trigger mode is designed to capture a signal waveform only once when the specified trigger event occurs. After capturing the signal, the oscilloscope holds the display, allowing for detailed analysis of that specific event.

Ideal for capturing rare, one-time events or anomalies that do not occur regularly, such as sudden spikes or drops in voltage.

Advantages	Disadvantages
• Provides a clear and stable view of the event, making it easier to analyse and diagnose specific issues without the interference of continuous data. • Can be used to capture a single event without having to look at the oscilloscope screen.	• It requires manual reset or re-arming to capture subsequent events, which can be time-consuming if multiple events need to be analysed.

Pulse width trigger

A Pulse Width Trigger (if available) is used to capture signals based on the duration of their pulses. It allows the oscilloscope to detect variations in pulse widths, which can be crucial for diagnosing issues related to timing and pulse duration in automotive systems. For instance, it can help identify problems in fuel injection systems where precise pulse width control is necessary.

Runt trigger

A Runt Trigger (if available) captures signals that fail to reach a full amplitude, often indicating irregularities or faults in the circuitry. Runt signals typically have low amplitude and can be symptomatic of issues like poor connections, intermittent faults, or degraded components in an automotive system. By setting up a runt trigger, you can isolate and identify these weak signals for further analysis.

Oscilloscope Technology and Features

Pattern Recognition trigger

Pattern Recognition Triggers (if available) are used to detect specific sequences or patterns within a signal. This advanced trigger type is particularly useful for identifying complex or recurring issues in automotive electronics. For example, it can recognise patterns in engine control signals or communication protocols like CAN or LIN.

Using triggers on a digital storage oscilloscope

Setup and use

Step 1
- Power On the Oscilloscope: Ensure the oscilloscope is connected to a power source and turn it on.

Step 2
- Connect Probes: Attach the oscilloscope probes to the points in the vehicle's electrical system that you wish to analyse.

Step 3
- Select the Trigger Menu: Access the trigger menu on the oscilloscope. This is usually done via a dedicated button or through the on-screen interface menu. Choose a simple trigger type (atuo, repeat, single) or configure more advanced triggering.

Step 4
- Choose the Trigger Type: If available/required select the advanced trigger type – pulse width, runt, or pattern recognition – from the available options. These can sometimes be found in the measurement section of the oscilloscope tools.

Step 5
- Set the Trigger Parameters: For Pulse Width Trigger - define the pulse width range by setting the minimum and maximum duration values. For Runt Trigger - specify the amplitude threshold that signals must cross to be considered a runt signal. For Pattern Recognition Trigger - define the specific pattern, sequence, or logic conditions that the oscilloscope should recognise.

Step 6
- Adjust the Trigger Level: Use the trigger control to set the voltage level at which the oscilloscope will trigger. This ensures that the oscilloscope captures the desired events accurately.

Step 7
- Enable the Trigger: Activate the trigger mode, ensuring that the oscilloscope is ready to capture signals based on the defined criteria. Decide whether it should trigger on a rising or falling voltage.

Step 8
- Analyse the Captured Signals: Operate the system or circuit, and once the oscilloscope triggers on the specified event, analyse the displayed waveform for diagnostics and troubleshooting.

Step 9
- Save and Document Findings: Save the captured waveforms and document your observations for future reference.

Oscilloscope Technology and Features

 If a simple trigger is used, try switching between repeat and auto, to see which provides the most useful capture.

High-Speed Sampling

High-speed sampling enables the capture and analysis of intricate details found in rapidly changing signals. By understanding the impact of sampling rates and resolution on waveforms, you can ensure accurate diagnostics and troubleshooting of automotive systems.

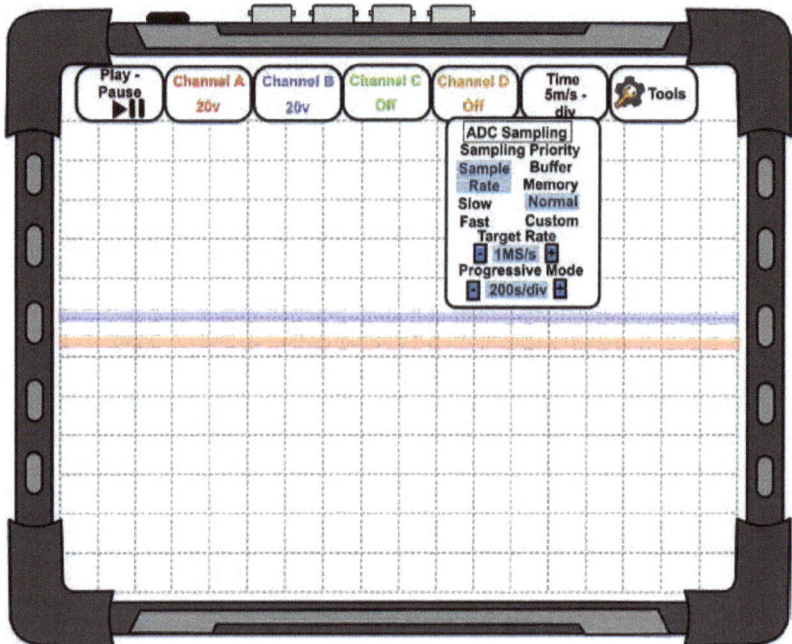

Figure 4.3 Sampling menu

Impact of sampling rates on waveforms

Sampling rate, measured in samples per second (SPS), dictates how frequently the oscilloscope captures data points from a signal. A higher sampling rate allows for more data points per unit of time, resulting in a more detailed representation of the waveform. Conversely, a lower sampling rate may miss significant changes in the signal, leading to an incomplete or distorted waveform.

 The Swedish-American engineer and physicist Harry Nyquist known for his contributions to telecommunications and signal processing, developed the Nyquist theorem. This says that in digital signal processing, the sampling rate must be at least twice the highest frequency of a signal to avoid problems where high-frequency components of the signal are incorrectly represented as lower frequencies. This is known as '**aliasing**'.
When powered on, most oscilloscopes will default to a pre-set sampling rate. This can often be adjusted in small increments using + or – buttons.

Oscilloscope Technology and Features

Resolution refers to the number of **bits** used by the oscilloscope to represent each sample. Higher resolution allows for finer distinctions between voltage levels, leading to greater accuracy in the waveform representation.

- Bit Depth: Common resolutions include 8-bit, 12-bit, and 16-bit. An 8-bit **resolution** can represent 256 distinct levels, while a 16-bit resolution can represent 65,536 levels.
- Signal Fidelity: Increased resolution enhances **signal fidelity**, allowing you to observe subtle variations in the waveform that may indicate underlying issues.
- Noise Reduction: Higher resolution helps reduce noise, resulting in a cleaner and more accurate signal representation.

Aliasing - when a digital sensor or oscilloscope misinterprets a high-frequency signal as a lower-frequency one due to insufficient sampling rate, leading to distorted data.

Bits - the smallest units of digital data used in electronic control units (ECUs) and communication protocols, determining signal resolution and processing accuracy.

Resolution - the level of detail an oscilloscope can display, determined by its ADC bit depth, affecting signal accuracy and clarity in diagnostics.

Signal fidelity - the accuracy and integrity of a signal in electronic systems, ensuring minimal distortion or noise for precise diagnostics and control.

Setting up sampling speeds for accurate measurements

Setup

Step 1
- Determine the Signal Characteristics: Identify the highest frequency component of the signal you are measuring. This will influence the minimum required sampling rate based on the Nyquist theorem.

Step 2
- Select the Appropriate Sampling Rate: Choose a sampling rate at least twice the highest frequency of the signal. For complex signals, consider using a rate significantly higher than the minimum to capture transient events.

Step 3
- Configure the Resolution: Set the resolution based on the required accuracy and the nature of the signal. Higher resolution is preferable for detailed analysis and noise reduction.

Step 4
- Adjust the Memory Depth: Ensure the oscilloscope has sufficient memory to store the captured data. Longer signal captures and detailed analysis may require increased memory allocation.

Step 5
- Validate the Settings: Perform test measurements to verify that the chosen sampling rate and resolution accurately capture the signal without aliasing or loss of detail.

Oscilloscope Technology and Features

Memory Depth Management

Memory depth management is an essential concept you must grasp to maximise your oscilloscope's potential. This section outlines how to use memory effectively for prolonged signal captures and in-depth analysis.

Settings and terminology for memory depth differ among manufacturers. The description provided here is generic; it is advised to refer to your own oscilloscope operating manual and practice to understand the specific setup.

Understanding memory depth

Memory depth is the data an oscilloscope can record while acquiring a signal. More memory depth means more data points and detailed recordings, useful for diagnosing intermittent electrical issues by capturing subtle anomalies over time.

Setting up memory depth for long signal captures

Setup

Step 1
- Determine the signal capture duration based on your diagnostic needs. Calculate memory requirements using the signal frequency and sampling rate. High sampling rates over several seconds will need a lot of memory.

Step 2
- Set the oscilloscope's memory depth according to the calculated requirement. Many oscilloscopes will allow you to adjust the memory settings.

Step 3
- Optimise the time base setting to match the duration of your signal capture. A longer time base will stretch the signal over a wider range, allowing for detailed observation of longer captures.

Detailed analysis with enhanced memory depth

Use

Step 1
- Use the oscilloscope's zoom functions to examine specific parts of the captured signal. Focus on areas of interest to detect subtle waveform variations.

Step 2
- Apply advanced signal filtering techniques to eliminate noise and highlight important signals. Adjust filters to isolate frequency components or smooth out data.

Step 3
- Utilise tools like cursors, rulers, maths channels, or deep measure to identify issues and faults.

Oscilloscope Technology and Features

Specialised Setup Techniques

 While an oscilloscope can be a useful tool, it is not always suitable for diagnosing and testing high-voltage electric vehicle systems. Some oscilloscopes and attachments might not meet the necessary safety requirements for working on electric vehicles. If you do conduct tests, always observe all health and safety precautions, including the use of high-voltage Personal Protective Equipment (PPE).

You may also need additional probes, such as current loops or differential oscilloscope probes. Always follow the manufacturer's instructions.

Never conduct any testing on the high-voltage systems of a hybrid or electric vehicle unless you have received appropriate training and have the necessary experience.

Differential probes

A **differential probe** is an oscilloscope attachment that can measure the voltage difference between two points of an electric circuit, which can often be affected by an issue known as **common-mode voltage**. Common-mode voltage occurs when an identical component voltage is present at both terminals of an electrical device and can interfere with the signal or cause damage to the equipment.
It consists of two input leads, a **differential amplifier**, and an output lead, which is connected to the oscilloscope. The integrated differential amplifier subtracts the voltages at the two input leads and amplifies the difference.

Why differential probes are needed

In modern vehicles, especially hybrids and electric cars, high-voltage systems are becoming increasingly common. These systems can pose significant challenges for diagnostics due to their potential to generate noise and interference. Using differential probes allows you to safely and accurately measure voltages in these systems without the risk of damaging the equipment or misinterpreting the signals (as long as they conform to the correct safety specifications). They help in isolating noise and provide a clearer picture of the electrical behaviour within the vehicle's circuitry.

A differential probe is used to measure:

- High-voltage signals that exceed the **ground-referenced input range** of the oscilloscope.
- Low-voltage signals that are hidden by noise or interference.
- Differential signals used in communication or data transmission systems.
- **Floating** or isolated **signals**, similar to those generated in battery-powered devices or transformers.

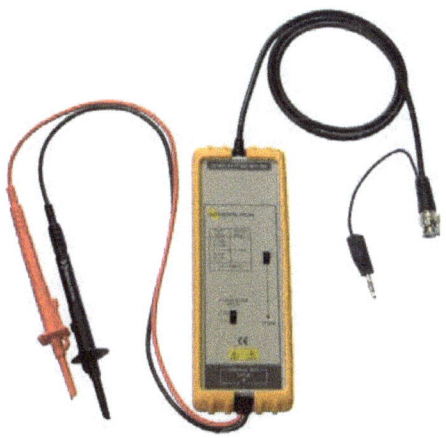

Figure 4.4 A differential probe

Oscilloscope Technology and Features

Differential probe - a specialised oscilloscope accessory that measures voltage differences between two points, essential for analysing high-voltage circuits and avoiding ground loops.

Common-mode voltage - a shared voltage that can interfere with signals, making it harder to get accurate readings from sensors and circuits.

Differential amplifier - a component that boosts the voltage difference between two input signals while rejecting common noise, ensuring accurate signal processing in sensors and electronic circuits.

Ground-referenced input range - the voltage limits that a sensor or circuit can accurately process when measured relative to the vehicle's electrical ground.

Floating signals - electrical signals that do not share a direct connection to the vehicle's ground, requiring special probes or isolation techniques for accurate measurement.

Differential probes are often category rated in a similar manner to that found on a multimeter. To be used on the high-voltage systems of a hybrid or electric car, they need to be rated at a minimum of CATIII 1000 volts.

Do not use a differential probe on an electric vehicle high-voltage system unless you have received appropriate training and are observing all relevant safety precautions and wearing the correct high-voltage Personal Protective Equipment (PPE).

How to use differential probes

Differential probes might seem daunting at first, but their setup and use can be straightforward with the right guidance.

Guide to setting up and using differential probes

Setup and use

Step 1
- Select the Appropriate Probe: Choose a differential probe that matches the voltage range of the system you are diagnosing. Ensure it is compatible with your oscilloscope.

Step 2
- Safety Considerations: Always ensure that you are working in a safe environment when dealing with high-voltage systems. Proper insulation, grounding, and personal protective equipment are essential.

Step 3
- Connect the Probe to the Oscilloscope: Attach the differential probe to the oscilloscope's input channel(s).

Oscilloscope Technology and Features

Step 4
- Attach the Probe Leads: Most differential probes have two leads: positive (+) and negative (-). Connect the probe leads to the points in the circuit you wish to measure. Ensure the connections are secure to avoid inaccurate readings.

Step 5
- Adjust Oscilloscope Settings: Set the oscilloscope to differential mode and adjust the voltage scale to match your probe's specifications. This ensures accurate representation of the signal.

Step 6
- Calibrate the Probe: Perform a calibration to ensure the probe is accurately measuring the voltages. Follow the manufacturer's instructions for any calibration procedures.

Step 7
- Capture and Analyse the Signal: Use the oscilloscope to capture the signal from the differential probe. Look for any anomalies or unusual readings that could indicate issues within the high-voltage system.

Isolation techniques

When working with hybrid and electric vehicles, you must be acutely aware of the dangers associated with high-voltage systems. Isolation techniques are paramount in ensuring the safety of both the technician and the vehicle.

- Proper Insulation: Using insulation materials and tools designed for high-voltage applications is crucial. Ensure that all tools and equipment are rated for the voltage levels you are working with. Insulated gloves, mats, and covers provide an added layer of protection.
- Disconnecting High-Voltage Systems: Before starting any work, always disconnect the high-voltage battery. Most hybrid and electric vehicles have a service disconnect switch or plug, which should be used to safely isolate the high-voltage system.
- Personal Protective Equipment (PPE): Wearing appropriate PPE, such as high-voltage-rated gloves, face shields, and protective clothing, is essential. These items are designed to protect against electrical arcs and shocks.
- Isolation Barriers: Physical barriers, safety cains, plexiglass shields, or insulated covers can be used to isolate high-voltage areas and components during diagnostics and repairs. This minimises the risk of accidental contact.
- Regular Training and Certification: Ensure that you are up-to-date with the latest safety protocols and techniques by undergoing regular training and certification. This knowledge is crucial in mitigating the risks associated with high-voltage systems.

Systematic high-voltage battery isolation approach

In order to shut down or isolate the high-voltage system of a hybrid or electric vehicle, manufacturers' procedures should always be adhered to. However, the following approach can be used as a systematic procedure to help ensure safety.

Figure 4.5 Isolating a high-voltage electric vehicle

Oscilloscope Technology and Features

Setup

Step 1
- To ensure everybody knows that work is being conducted, signs and barriers should be placed around the vehicle, creating a buffer zone to help keep people away. This buffer zone should extend beyond arm's length of others.

Step 2
- The vehicle should be powered up and placed in ready mode to ensure that no malfunction indicator lights are present, which may show a fault with the vehicle. The vehicle should then be switched-off, key removed and placed beyond its range of operation if it is a smart key.

Step 3
- If accessible, the negative terminal of the 12 volt auxiliary battery should be disconnected and isolated so that it cannot accidentally reconnect. *(See diagnostic tip).*

Step 4
- Before any high-voltage isolators or components are touched, high-voltage PPE including Class 0 gloves and a face shield should be inspected, tested and worn.

Step 5
- Following manufacturers' instructions, remove the high-voltage isolator, maintenance service disconnect (MSD) and place it in a secure location so that it cannot accidentally be reconnected. If the vehicle has a high-voltage interlock loop (HVIL), this should be disabled and secured using a lockout to ensure that it cannot accidentally be reconnected.

Step 6
- The operator should then wait the recommended time to allow capacitors to discharge.

Step 7
- Following manufacturers' instructions, a calibrated multimeter should then be used to check for absence of voltage from the high-voltage battery and capacitors.

Before disconnecting the low-voltage auxiliary battery, conduct a voltage test with a voltmeter; this will confirm several things:

- The voltmeter is working, accurate, and set to the correct measurement unit.
- The vehicle has shut down and switched off the DC-to-DC converter. (If charging voltage is seen, the high-voltage system is still awake. Do not disconnect the low voltage auxiliary battery.)

An additional precaution that can be used during the shutdown and isolation procedure is to try and place the vehicle in ready mode again after the smart key has been placed beyond its range of operation. This will normally show 'Key not detected' on the driver's display. This helps to prove that:

- The vehicle is switched off.
- The key is beyond its range of operation.
- There is no spare key located inside the vehicle.

Oscilloscope Technology and Features

Insulated tools and their importance in relation to short circuit

It is essential to use correctly rated and insulated hand tools when working on or around the high-voltage systems of an electric vehicle. The classification of these tools is normally conducted by the International Electrotechnical Commission (IEC), which is an organisation that publishes the standards for electrical and electronic related technologies. These tools should be classified and tested to IEC 60900, which is applicable to insulated, insulating and hybrid hand tools used for working live or close to live parts at nominal voltages up to 1000 V AC and 1500 V DC. These tools can often be identified by a 1000 V (double triangle) IEC EN 60900 mark with the name of the manufacturer, the tool reference, and the year of manufacture.

Some insulated tools may be classified under the German VDE standard Verband der Elektrotechnik, Elektronik und Informationstechnik (Association for Electrical, Electronic and Information Technologies).

The main purpose of insulated tools is to reduce the possibility of electrical short circuit when working on or around the high-energy electrical systems. A short circuit may cause arcing, rapid heating, fire, or even explosion, leading to injury or death.

The insulation performs a secondary function, which can be an additional protection to the user or operator against electrocution. However, insulated tools should always be used in addition to high-voltage electrical Personal Protective Equipment (PPE) and never instead of.

Figure 4.6 Insulated hand tools

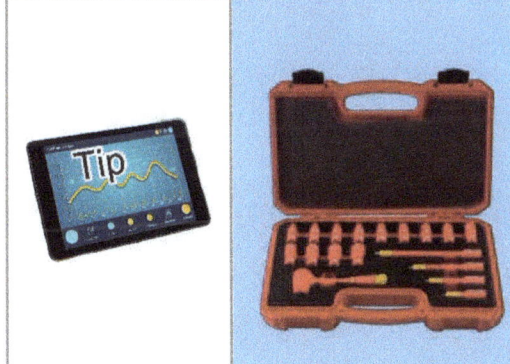

Although the colour coding of insulated hand tools is not standardised, they are often coated in red or orange coverings to indicate their potential use with high-voltage systems.

Some manufacturers provide a double coating for their tooling, first in yellow and then covered with red insulation. This means that the user or operator can easily identify any damaged or worn-out tooling. If the tool wears or is damaged, the yellow under insulation is exposed.

Following these isolation techniques can significantly reduce the risk of electrical hazards while working on hybrid and electric vehicles. Always prioritise safety and adhere strictly to manufacturer guidelines and industry standards.

Oscilloscope Technology and Features

Custom probe calibration

When diagnosing electrical faults, precise and reliable signal measurement is essential. Oscilloscopes are used to capture and display electrical signals, and their probes must be correctly calibrated for accurate measurements. Calibration involves adjusting the probe settings to align with the oscilloscope's input impedance and the specific requirements of the measurement task. This process is particularly important when using probes and scopes from different manufacturers.

Probe calibration requires knowledge of the signal type to be measured. Different signals may need various attenuation settings, such as 1x, 10x, or higher. Choosing the correct attenuation factor is important to ensure the signal is properly represented on the oscilloscope display.

Once the attenuation factor is set, the next step involves compensation. Compensation adjusts the probe to match the input capacitance of the oscilloscope. This is done by connecting the probe to a known reference signal (some oscilloscopes have built-in calibration signals).
In addition to attenuation and compensation, the physical condition of the probe must be considered. Damage or wear to the probe's insulation or connectors can affect signal integrity. Regular inspection and maintenance of probes are necessary to maintain their accuracy and reliability.

Modern oscilloscopes, like the ones provided by industry leaders, offer facilities that allow you to design and configure measurement parameters for custom or unusual probes. This flexibility is useful for specific diagnostic needs or when standard probes do not meet your requirements.

A key benefit of custom probe configuration is the ability to adjust the probe characteristics according to the specific needs of the measurement environment. This involves modifying the attenuation, compensation, and physical design of the probe to achieve optimal performance and accuracy.

Setting up a custom probe starts with defining the electrical characteristics to measure, including voltages, frequencies, and signal properties. The display can be labelled with output units like pressure and temperature. After establishing these parameters, input the settings into the oscilloscope software to match the probe to the oscilloscope's capabilities.

Some oscilloscope software can create custom calibration profiles to store unique settings for each probe. Once configured, these profiles can be saved and recalled for consistent and accurate measurements.

Figure 4.7 Custom probes

Advanced signal filtering

Noise and interference often reduce signal measurement accuracy. Advanced signal filtering improves the clarity and precision of signals captured by oscilloscopes and can help you diagnose electrical faults effectively.

Understanding noise and its impact

Noise in electrical signals can come from various sources, including electromagnetic interference (EMI), power supply fluctuations, and the inherent characteristics of the electronic components being measured. This noise can interfere with the actual signal, complicating the identification of the underlying issue. The use of filters to eliminate noise can be beneficial for acquiring precise and reliable data.

Oscilloscope Technology and Features

Types of filters and their applications

Different types of filters can be used to clean up signals, each with its specific application and benefits:

Low-Pass Filters:

- These filters allow signals below a certain frequency to pass through while smoothing out higher frequency noise. They are particularly useful for filtering out high-frequency interference in slow-changing signals.

High-Pass Filters:

- These filters allow signals above a certain frequency to pass through, removing low-frequency noise. They are ideal for reducing power supply fluctuations and other low-frequency disturbances.

The process of applying filters typically involves selecting the appropriate filter type and configuring its parameters, such as cutoff frequency and bandwidth.

Applying filters

Use

Step 1
- Select the Filter Type: Based on the characteristics of the noise and the signal of interest, choose the appropriate filter type (low-pass, high-pass).

Step 2
- Configure Filter Parameters: Set the cutoff frequency, bandwidth, and other relevant parameters to match the specific requirements of the measurement task. Some oscilloscope software will have an up/down or +/- button that you can simply click to alter the filter. This is a simple way to fine-tune the capture.

Step 3
- Apply the Filter: Use the oscilloscope software to apply the filter to the captured signal. Observe the filtered signal and verify that the noise has been effectively removed while the critical components remain highlighted.

Benefits of advanced signal filtering

Using advanced signal filtering techniques offers several advantages:

- Enhanced Signal Clarity: Filters remove unwanted noise, providing a clearer and more accurate representation of the true signal.
- Improved Diagnostic Accuracy: With cleaner signals, you can more easily identify and diagnose electrical faults.
- Reduced Time and Effort: Filtering simplifies the analysis process, saving time and reducing the effort required to interpret complex signals.
- Customisable Filtering Options: Oscilloscope software allows for the customisation of filter settings, enabling you to tailor the filtering process to specific diagnostic needs.

Oscilloscope Technology and Features

 It is important to note that filtering, particularly low-pass filtering, can alter or reduce the physical shape of the waveform. Therefore, it should always be compared and used alongside the original image.

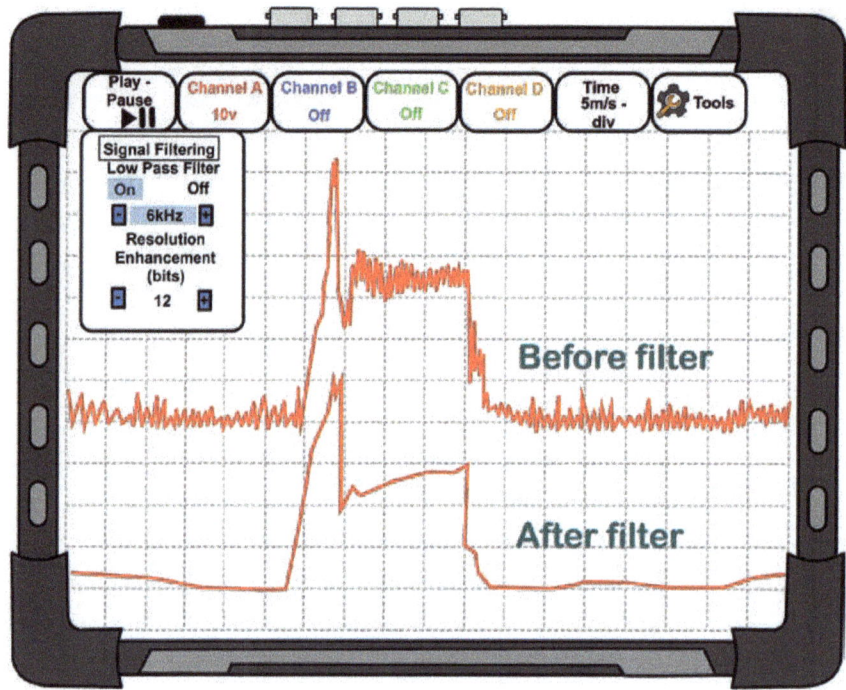

Figure 4.7 Signal filtering

 Think of low-pass and high-pass filtering like adjusting the bass and treble on a car's audio system.

- Low-Pass Filter: This is like turning up the bass and reducing the treble. It allows the deep, slow-moving sounds (low-frequency signals) to come through while filtering out the sharp, high-pitched noises (high-frequency interference). In an oscilloscope, this helps remove high-frequency noise, making it easier to see slower signals like temperature sensor readings.
- High-Pass Filter: This is like boosting the treble while cutting the bass. It lets the sharp, fast-moving sounds (high-frequency signals) come through while filtering out the slow, rumbling tones (low-frequency noise). On an oscilloscope, this helps eliminate unwanted low-frequency interference, making it easier to analyse fast-changing signals like ignition coil pulses.

Oscilloscope Technology and Features

Integration with Other Tools

Combining oscilloscopes with diagnostic scan tools

In an automotive repair environment, leveraging multiple diagnostic tools can significantly enhance the accuracy and efficiency of fault identification and resolution. Combining oscilloscopes with diagnostic scan tools is a powerful method that allows you to blend real-time waveform analysis with robust diagnostic data, providing a holistic view of the vehicle's electrical and electronic systems.

Advantages

- Comprehensive Data Analysis: By integrating the detailed waveforms captured by oscilloscopes with the diagnostic trouble codes (DTCs) and live data obtained from scan tools, a more thorough understanding of the issues can be achieved. This synergy enables the correlation of electrical behaviours with specific diagnostic codes.
- Fault Isolation: Oscilloscopes can detect subtle electrical anomalies, while scan tools offer broader system information. Using both tools together helps identify the exact location and nature of faults, reducing guesswork and improving repair precision.
- Improved Efficiency: Combining tools speeds up diagnostics by quickly verifying scan tool data with oscilloscope waveforms, saving time and streamlining troubleshooting.
- Informed Decision Making: Combining oscilloscope readings with scan tool data provides a thorough dataset, enabling more accurate repair decisions. This can lead to lower misdiagnosis rates and fewer unnecessary part replacements.

Disadvantages

- Complexity of Integration: Combining data from oscilloscopes and scan tools requires knowledge of both instrument types and their software interfaces. Synchronising data from multiple sources can be challenging and time-consuming.
- Cost Implications: Smaller workshops may struggle with the expense of high-quality oscilloscopes and diagnostic scan tools.
- Training Requirements: The effective use of combined tools often requires comprehensive training. Technicians need to be skilled in interpreting oscilloscope waveforms and scan tool data, which necessitates ongoing learning and staying updated with technological advancements.

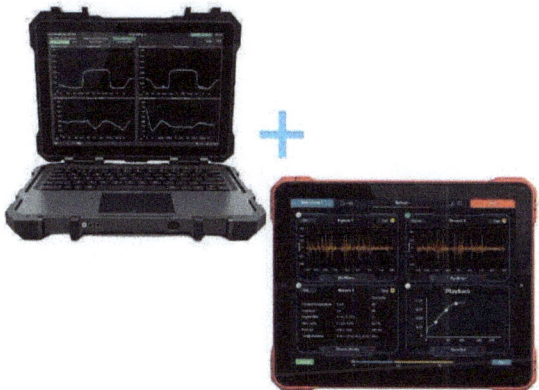

Figure 4.8 Oscilloscopes and scan tools

 Scan tool graphs should not be mistaken for oscilloscope waveforms. While both display signal behaviour, scan tool graphs are derived from processed serial data reported by the vehicle's control modules, meaning they show second-hand, time-delayed values that have already been interpreted by the ECU. In contrast, oscilloscope waveforms provide real-time, raw electrical measurements directly from the circuit, capturing live voltage changes without ECU filtering.

Oscilloscope Technology and Features

Using oscilloscopes with multimeters

Using oscilloscopes together with multimeters allows you to analyse automotive electrical systems comprehensively. These tools complement each other by offering a combination of broad and detailed perspectives on system behaviours.

Advantages

- Comprehensive Data Analysis: Integrating the high-resolution readings of oscilloscopes with the precise measurements of multimeters enables a thorough understanding of electrical issues. Oscilloscopes capture detailed waveforms while multimeters provide accurate voltage, current, and resistance readings. This combination allows for the correlation of real-time electrical behaviours with specific measurements, thereby improving diagnostic precision.
- Improved Fault Isolation: Together, these tools help in pinpointing the exact location and nature of faults, reducing guesswork, and improving repair accuracy. This combination is particularly effective in identifying intermittent issues that might be overlooked using a single tool.
- Informed Decision Making: Integrating oscilloscope waveforms with multimeter measurements provides a comprehensive dataset. This integration enables you to make well-informed and confident repair decisions, leading to lower misdiagnosis rates and fewer unnecessary part replacements. The detailed insights obtained from both tools ensure repairs are grounded in accurate data, thereby enhancing the overall quality and reliability of your services.
- Enhanced Troubleshooting Efficiency: Using oscilloscopes and multimeters together improves the diagnostic process. you can quickly verify multimeter readings with oscilloscope waveforms, reducing the time needed for data cross-referencing and speeding up troubleshooting. This integrated method reduces the need for repeated tests and rechecks, making the repair process faster.

Disadvantages

- Complexity of Integration: Combining data from oscilloscopes and multimeters requires knowledge of both instrument types and their respective interfaces. you need to synchronise data from multiple sources, which can be difficult and take time. Interpreting and correlating the data from these tools demands expertise.
- Cost Implications: Investing in high-quality oscilloscopes and multimeters can be financially challenging, particularly for smaller workshops. The expenses associated with acquiring and maintaining advanced diagnostic equipment may limit accessibility to these technologies for some technicians.
- Training Requirements: Effective use of combined tools often requires comprehensive training. Technicians must be proficient in interpreting oscilloscope waveforms and multimeter readings, which involves continuous learning and staying updated with technological advancements. This ongoing education demands time and resources, which may be challenging for some professionals to manage.

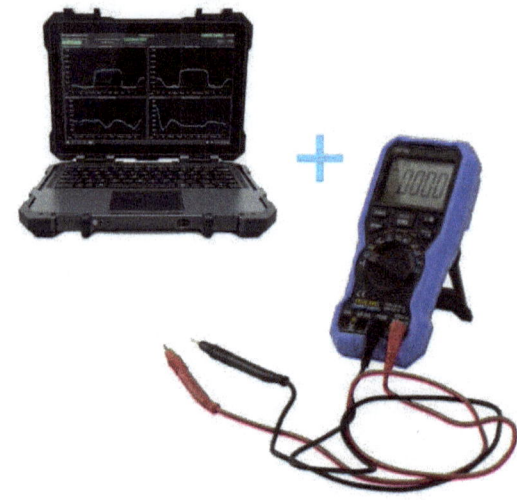

Figure 4.9 Oscilloscopes and multimeters

Oscilloscope Technology and Features

Think of automotive diagnostics like solving a puzzle with different tools.

- The scan tool is like reading a detective's case file—it tells you what the car's ECU has observed, reporting stored trouble codes and system data. But it only shows symptoms, not the root cause.

- The multimeter is like a magnifying glass—it allows you to check voltage, resistance, and continuity in individual circuits, helping pinpoint basic electrical faults.

- The oscilloscope is like a high-speed camera—it captures electrical activity in real-time, revealing dynamic signal behaviours such as injector pulse patterns, ignition coil firing, and sensor waveforms.

Using all three together transforms troubleshooting from guesswork into precision diagnosis. The scan tool provides the initial clue, the multimeter confirms electrical health, and the oscilloscope uncovers hidden faults that cause intermittent issues.

Although there are some challenges, integrating oscilloscopes with other electrical test tools presents a notable improvement in automotive diagnostics. By leveraging the capabilities of multiple instruments, you can achieve high accuracy and efficiency when diagnosing and repairing complex electrical issues.

Conclusion

Oscilloscopes differ based on the manufacturer, and their designs will vary accordingly. As technology develops, features are likely to increase and become more automated, facilitating quicker and more efficient diagnosis. Remember, however, it is the technician's skill that drives the diagnosis and repair process. Regular practice with an oscilloscope is essential for becoming proficient in its use.

Chapter 5 Mastering Signal Analysis

Signal analysis is a necessary skill for technicians. As vehicles become more complex, the ability to understand electronic signals can distinguish between an accurate diagnosis and a missed issue. This chapter covers the techniques and tools needed to analyse signals using automotive oscilloscopes.

You will learn to identify signal anomalies such as glitches, spikes, and dropouts that are important for finding problems. Annotating waveforms and overlaying them with reference signals can offer deeper insights into the vehicle's electronic condition.

Additionally, this chapter provides guidance on vehicle networks, enabling the capture and interpretation of communication protocols.

Contents

Waveform Libraries	**115**
Guided Tests and Presets	**118**
Waveform Analysis	**119**
Annotating and Overlays	**121**
Timing Measurements	**125**
Pulse Width Diagnostics	**126**
CAN BUS and In-vehicle Networks	**133**
Application-Specific Diagnostics	**144**

The automotive industry is a high-risk environment, especially when dealing with electrical systems. The hazards of electricity are well-known but can be easily ignored due to its invisible nature. This can lead to complacency if the fundamentals of electricity are not well understood. Even with this understanding, caution is necessary. Assume that any safety systems designed for protection have failed and take precautions to minimise the risk of injury or death. Always evaluate the risks associated with any activity and implement measures to eliminate or reduce the hazards involved in any task, diagnosis, or repair. Additional risks associated with working on, or around electrical systems may include:

- Electrocution
- Strong magnetic fields
- Falling from heights
- Short circuits
- Electrical discharge/arcing
- Fire and explosion
- Chemicals

Mastering Signal Analysis

Waveform Libraries

In automotive diagnostics, precision and accuracy are essential. Oscilloscopes have become indispensable tools, allowing you to visualise complex electrical signals and diagnose issues with a high degree of accuracy. A crucial element that enhances the effectiveness of this diagnostic tool is the availability of reference waveforms and **waveform libraries**. These libraries contain '**known good**' waveform examples that serve as benchmarks for comparison during diagnosis.

Why access to 'known good' waveform examples is crucial

When diagnosing automotive systems, having access to known good waveform examples is indispensable. These waveforms represent the expected signals from correctly functioning components and systems. Here are several reasons why they are essential:

- Establishing a Baseline: Known good waveforms provide a baseline against which current measurements can be compared. By knowing what a healthy signal should look like, you can quickly identify deviations that indicate potential issues. This comparison is crucial for accurate diagnostics and helps in pinpointing the exact nature of faults.
- Time Efficiency: Having access to a library of known good waveforms can significantly reduce the time required for diagnosis. Instead of starting from scratch, you can refer to these examples to quickly identify discrepancies. This efficiency is particularly beneficial in an environment where time is of the essence.
- Minimising Misdiagnosis: Misdiagnosis can lead to unnecessary repairs and increased costs. Known good waveforms act as a reliable reference, minimising the chances of misinterpreting signals. By ensuring that the diagnosis is based on accurate comparisons, the likelihood of errors is greatly reduced.
- Training and Education: For new mechanics or those unfamiliar with specific systems, known good waveforms serve as an educational tool. They help in understanding the expected behaviour of electrical signals and improve the diagnostic skills of the user.

Waveform libraries - a collection of pre-recorded reference waveforms used for diagnosing and analysing vehicle electrical and sensor signals.

Known good - a verified, correctly functioning component, circuit, or waveform used as a reference for diagnostics.

Reference waveforms

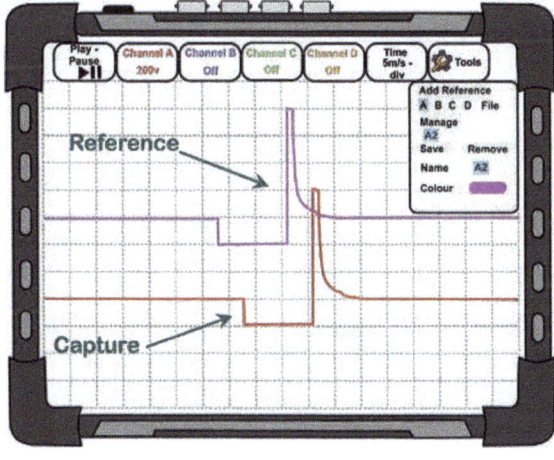

Reference waveforms are a handy feature offered by some oscilloscope manufacturers. They allow you to capture a correctly operating waveform example and display it on the screen as an overlay. Then, as long as the same operating conditions exist with the circuit or component under test (including time base and amplitude settings) a direct comparison can be made to help identify any anomalies. Known good reference waveforms can often be labelled and stored for future use, creating your own waveform library.

Figure 5.1 Reference waveforms

Mastering Signal Analysis

How to set up and store reference waveforms

Setup and use

Step 1: Set up the oscilloscope to capture an example of a known good waveform. This could be from a correctly operating circuit on the vehicle under test, or from another vehicle which has no issues.

Step 2: Navigate to the reference waveform menu and select the channel that you want to use as the example. This can be done while the oscilloscope is running, active and it will hold a snapshot or the capture can be paused beforehand to ensure that you are happy with the example.

Step 3: The reference waveform is normally then set as a direct overlay of the example captured.

Step 4: Now set the oscilloscope up to capture the suspected faulty circuit or component, and operate so a live waveform is generated.

Step 5: The reference waveform can normally be positioned on the screen to the operators preference as an offset or directly overlay the running waveform. This is normally achieved using position buttons for offset and dely, or using a drag and drop function.

Step 6: A direct comparison can now be made with the reference waveform to help identify any anomalies and faults.

Step 7: When complete, the reference waveform can be deleted, or lablled with a title and operating values to be stored for future use and comparisons. This method will help build a waveform library of known good exmples.

Accessing and creating your own waveform libraries

Building a comprehensive waveform library is an ongoing process that involves both accessing existing resources and creating new ones.

Accessing existing libraries

Many automotive oscilloscope manufacturers provide access to extensive waveform libraries. These resources can often be accessed through:

- Manufacturer's websites
- Diagnostic software suites
- Technical support forums

These libraries are a valuable starting point and can be supplemented with additional examples as you encounter new scenarios.

Mastering Signal Analysis

Creating your own waveform library

To create your own waveform library:
- Capture Signals: Use your oscilloscope to capture waveforms from known good vehicles. Ensure that the vehicle is in optimal working condition to obtain accurate signals.
- Annotate and Save: Annotate the waveforms with relevant details such as the vehicle make and model, component tested, and the conditions under which the signal was captured. Save these annotated waveforms in an organised manner for easy retrieval.
- Regular Updates: Continuously update your library with new waveforms and scenarios. This practice ensures that your library stays relevant and comprehensive.

When using an oscilloscope for diagnostics, capture reference waveforms. Save them quickly to a folder for easy access later. While the oscilloscope is on, test nearby components, even if they're not the main focus, to get known good examples. Use a simple naming convention for saved waveforms, including system and component names, but keep vehicle details in annotated notes (e.g., ignition_coil_on-plug).

The role of internet forums and social media groups

The automotive diagnostic community is vast and active online. Joining internet forums and social media groups dedicated to automotive diagnostics can be an excellent way to obtain reference waveforms. **Table 5.1** describes how these platforms can be beneficial:

Table 5.1 Internet forums and social media

Community knowledge sharing	Forums and social media groups are platforms where mechanics from around the world share their experiences and knowledge. By participating in these communities, you can gain access to a wealth of information, including rare or unusual waveform examples that may not be readily available elsewhere.
Real-world scenarios	The waveforms shared on these platforms often come from real-world diagnostic scenarios. These examples can provide insights into how different issues manifest in waveforms, enhancing your understanding and diagnostic capabilities.
Networking and collaboration	Engaging with peers online allows you to network and collaborate with other professionals. This interaction can lead to valuable exchanges of information and the development of a more robust waveform library.

While the internet is a valuable resource, it is important to exercise caution.
Not all waveforms shared online are reliable. Here are some tips to avoid being misled:
- Verify the Source: Ensure that the waveforms come from credible sources. Look for posts from experienced professionals or reputable forums.
- Cross-Reference: Cross-reference the waveforms with other trusted sources before relying on them for diagnosis.
- Beware of Faulty Conditions: Be mindful that some waveforms may depict faulty conditions or operations. These should not be confused with 'known good' waveforms. Issues are too numerous to show all potential faults with a system or circuit, so always verify the normal operating conditions.

Mastering Signal Analysis

Guided Tests and Presets

Technicians and mechanics can benefit greatly from guided tests and presets offered by some oscilloscope manufacturers. These features are designed to streamline the diagnostic process, making it easier to set up the oscilloscope and begin testing with confidence.

Guided tests

Guided tests provide comprehensive instructions for setting up and using the oscilloscope to diagnose specific automotive systems or components. These guides often include step-by-step procedures, detailing everything from safety warnings, connecting the oscilloscope to the vehicle, to interpreting the waveforms that are generated during testing.

How guided tests help:

- They simplify the setup process, allowing even less experienced users to perform advanced diagnostics.
- They provide clear, concise instructions that reduce the likelihood of errors during setup and testing.
- They often include examples of expected results, making it easier to identify anomalies in the waveforms.

Example:
A guided test might show you how to connect your oscilloscope to a vehicle's fuel injector system. It would explain which leads to connect, what settings to use, and what a normal injector waveform should look like.

Presets

Presets are another helpful feature offered by some oscilloscope manufacturers. With presets, users can select the system or component they wish to test from a menu, and the oscilloscope will automatically adjust settings such as time base, amplitude, and sometimes even cursors and rulers. This can save a significant amount of time and ensure that the oscilloscope is configured correctly for the specific test.

How presets help:

- They provide a quick and easy way to set up the oscilloscope for specific tests.
- They ensure that the settings are appropriate for the system or component being tested.
- They help users get accurate readings without having to manually configure each setting.

Example:
If you want to test an alternator, you would select 'alternator' from the preset menu. The oscilloscope will then adjust its settings to those optimal for capturing alternator waveforms, saving you the hassle of manual configuration.

While guided tests and presets are incredibly useful, it is important to remember that they should not be relied upon solely. Presets, in particular, may not always match the specific design or operating values of every manufacturer's system.

Beware:
- Presets might not cover all variations and nuances of different vehicle systems.
- Always cross-reference the results with manufacturer specifications and other reliable sources.
- Use presets as a starting point but verify the settings and results for accuracy.

Mastering Signal Analysis

Waveform Analysis

Waveform analysis is a fundamental skill for any technician or mechanic working with automotive oscilloscopes. Understanding how to identify signal anomalies such as glitches, spikes, and dropouts can elevate your diagnostic capabilities and ensure more accurate results. This section aims to provide clear and simplified explanations, making it accessible for all oscilloscope users.

What is a waveform?

A waveform is a graphical representation of a signal's voltage over time. By examining waveforms, you can gain insights into the performance and behaviour of various automotive systems and components. Oscilloscopes are invaluable tools in capturing these waveforms and allowing detailed analysis.

Identifying signal anomalies

Signal anomalies are irregularities or deviations in the expected waveform pattern. These can indicate underlying issues in the vehicle's electrical systems. Key types of signal anomalies are described in **Table 5.2**.

Table 5.2 Types of signal anomalies

Anomaly	Description
Glitches	A glitch is a brief, unexpected deviation from the normal waveform. • Appearance: Glitches typically appear as small, sharp spikes or dips in the waveform. • Causes: They can be caused by transient electrical noise, poor connections, or malfunctioning components. • Identification: To spot glitches, carefully inspect the waveform for any abrupt changes that are inconsistent with the expected pattern.
Spikes	Spikes are sudden, high-amplitude variations in the waveform. • Appearance: They appear as tall, narrow peaks or troughs. • Causes: Spikes can be caused by inductive kickback, switching transients, or electrical interference. • Identification: Look for significant, isolated peaks that stand out from the rest of the waveform. They often occur during transitions or switching events.
Dropouts	Dropouts are moments where the signal temporarily disappears or falls to zero. • Appearance: They appear as gaps or flat sections in the waveform. • Causes: Dropouts can be caused by intermittent connections, component failures, or disruptions in the signal path. • Identification: Scan the waveform for sections where the signal abruptly ceases or significantly reduces in amplitude.

Identifying dropouts and glitches using maths channels

Spotting dropouts or glitches in a waveform with large amounts of data can be challenging and time consuming. A quick method that can be used to help identify issues is to use a maths channel to measure frequency.

Setup and use

Step 1
- Connect the oscilloscope probe (e.g., Channel A) to the signal you want to analyse. Attach the ground to a suitable signal or chassis ground.
- Voltage Range: Set the input channel amplitude to a suitable range.

Step 2
- Time base: Set a time base that captures multiple cycles of the signal.
- Triggering: Use auto or rising edge trigger on the main signal to stabilise the waveform.

Step 3
- Create a Frequency Math Channel. To do this: Open Tools > Math Channels > Create. Use a predefined freq(ChannelA) function or type: freq(A).
- Set the unit to Hz and apply suitable smoothing or averaging if the scope allows. Colour-code it for visibility.

Step 4
- Analyse the Frequency Trace: The frequency line should be stable and flat during normal operation.
- Dropouts appear as sudden drops to 0 or near-zero frequency.
- Glitches may show as sharp spikes or sudden frequency changes, indicating unexpected fast or slow edges.

Step 5
- Use the zoom tools to examine sections of the frequency maths trace where the value deviates.
- Match up these deviations to the original waveform to confirm if:
- A dropout occurred (missing pulse, signal lost).
- A glitch occurred (brief invalid pulse or edge).

Step 6
- Compare Against Known Good: If available, load a known-good waveform with a frequency channel for the same component and compare behaviour.

Step 7
- Save and Document: Annotate areas with dropouts/glitches for future diagnosis or sharing with colleagues.

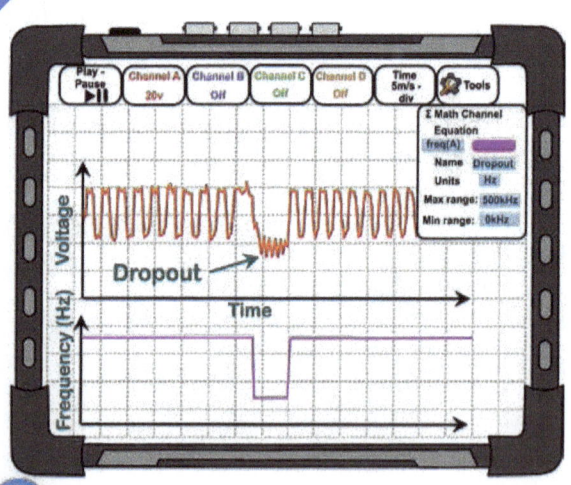

Figure 5.2 Maths channel for identifying dropouts or glitches

Mastering Signal Analysis

- Use low-pass filtering if your scope supports it, to reduce noise that may falsely trigger glitches.
- Ensure the signal you're analysing is clean; otherwise, frequency maths will reflect noise.
- Combine frequency maths with edge-counting measurements to see if dropouts are subtle.
- For signals like CAN, frequency maths can help detect bit stuffing or noise-induced retries.

In a CAN frame, bit stuffing ensures reliable communication by maintaining synchronisation.
Example:
Imagine transmitting the bit sequence 111110. Since there are five consecutive 1s, the protocol inserts a stuff bit (**0**) to create 1111100. This prevents excessive identical bits, ensuring proper edge transitions for synchronisation.
Similarly, if sending 000000, a stuff bit (**1**) is inserted after the fifth 0: 0000010.
Stuff bits are automatically removed by the receiver, restoring the original data.

Practical Tips for Waveform Analysis
- **Use Guided Tests and Presets:** Guided tests and presets on your oscilloscope can help set optimal parameters for capturing waveforms. However, remember to cross-reference with manufacturer specifications and verify the settings for accuracy.
- **Annotate and Overlay:** Use the annotation and any overlay features to mark anomalies and compare multiple waveforms. This can aid in pinpointing issues and understanding the context of the anomalies.
- **Timing Measurements:** Precise timing analysis is crucial for identifying synchronisation issues. Measure the timing of anomalies relative to other waveform events to diagnose potential causes. Reference waveforms can provide good examples for comparison.
- **Application-Specific Diagnostics:** Different automotive systems have unique waveform characteristics. Familiarise yourself with the typical waveforms of components like electric vehicle systems, injectors, ignition, and network communication systems to better identify anomalies.

Annotating and Overlays

Understanding how to effectively use your oscilloscope's features, such as annotations and overlays, can greatly enhance your diagnostic capabilities. These features allow you to mark important waveform components and compare different signals, making it easier to identify issues and understand their context. Many computer-based oscilloscopes allow mark-ups to be added to a waveform capture. This can be in the form of textboxes, arrows, markers, or if the equipment includes a touchscreen, drawings made directly on the image.

Mastering Signal Analysis

Understanding annotations

Annotations are notes or markers that you can place directly on your waveform to highlight specific points of interest. These might include anomalies, measurement points, or other significant events.
Here's how to use annotations effectively:

How to Use Annotations

- Identify Key Points: Mark the beginning and end of pulses, peaks, and troughs, or any other feature that stands out. This helps you and others quickly locate important parts of the waveform.
- Add Comments: Write brief notes explaining why a particular point is significant. For example, you might note that a spike in the waveform corresponds to a misfire in the engine.
- Use Different Markers: Some oscilloscopes allow you to use different symbols or colours for annotations. Use these to differentiate between various types of events or measurements.

Benefits of Annotations

- Clarity: Annotations make it easier to understand what you are looking at, especially when reviewing waveforms later or sharing them with colleagues.
- Efficiency: Quickly pinpointing important events saves time during diagnostics.
- Documentation: Annotated waveforms serve as excellent documentation for repairs and can be referenced in the future if similar issues arise.

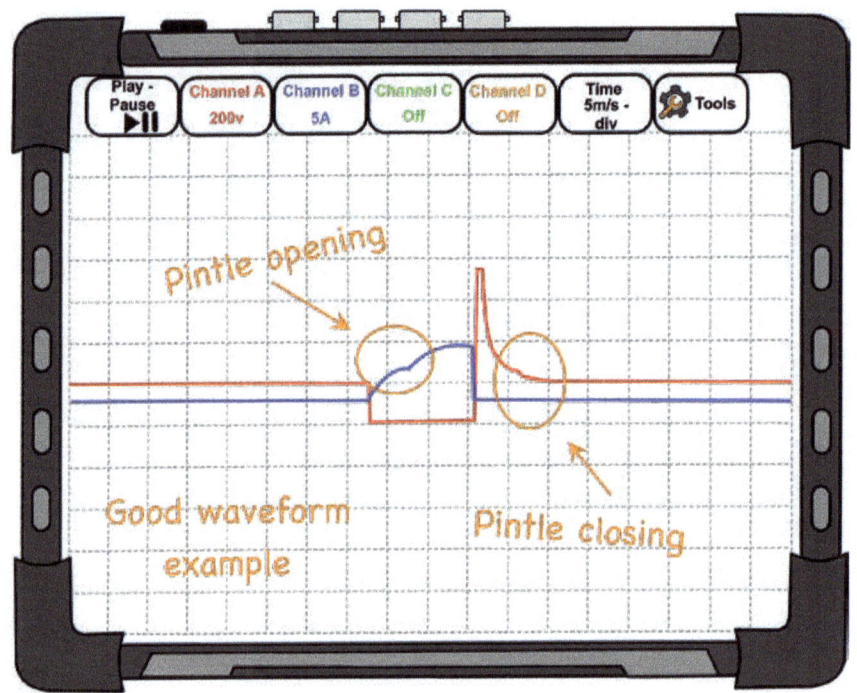

Figure 5.3 An annotated waveform

Mastering Signal Analysis

Tips for Using Annotations

- Renaming Channels: Always rename oscilloscope channels to reflect the signal being measured. For instance, instead of generic names like Channel 1, use descriptive names such as 'Injector Pulse' or 'Ignition Coil'. This clarity will greatly aid in future analysis and sharing.
- Adding Vehicle Details: Include relevant vehicle information within the annotations or notes. This should consist of the make, model, year, and specific engine details. This context is invaluable when comparing waveforms from different sessions or when collaborating with colleagues.
- Using the Notes Section: Oscilloscope software often comes with a dedicated notes section. Use this space to document your observations, hypothesis, and any corrective measures taken. Detailed notes can provide critical insights during troubleshooting and when revisiting old data.

Using overlays

Overlays allow you to display multiple waveforms on the same screen, making it easier to compare them. This feature is particularly useful when you need to see how different signals interact with each other or when comparing an unknown signal to a known good waveform.

Also, screen overlays are available using apps, or downloads which can be placed on top of a waveform capture and provide specific information such as firing order, engine cycles or other useful information.

How to use overlays

Setup and use

Step 1
- Select Waveforms: Choose the waveforms you want to overlay. These might be signals from different components or the same signal captured at different times.

Step 2
- Adjust Alignment: Align the waveforms so that key features coincide. This might involve shifting one waveform horizontally or vertically to match another.

Step 3
- Compare Features: Look for differences and similarities between the waveforms. Differences can help identify problems, while similarities can confirm that components are functioning correctly.

Step 4
- If identifying specific components or timings on a capture, download or launch a transparent overlay which can be placed over the waveform and adjusted for size and position. Use this to help understand events occurring within the capture, such as engine cycles for example.

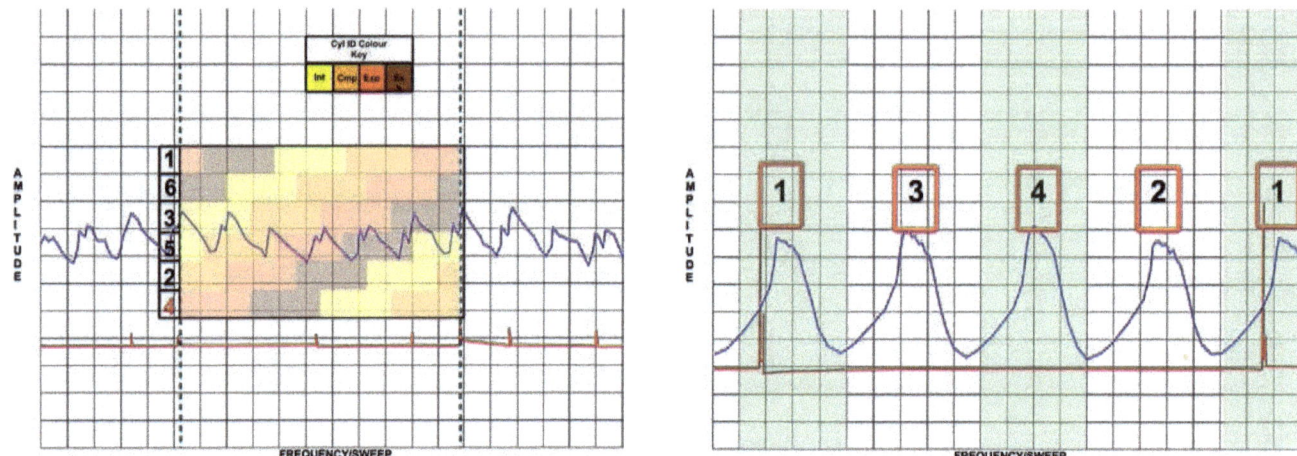

Figure 5.4 Waveform and screen overlays

Benefits of overlays

- Comparison: Easily spot discrepancies between expected and actual performance.
- Context: Understand how different signals relate to each other in real-time, and specific events that are occurring.
- Validation: Verify that repairs or adjustments have had the desired effect by comparing before and after waveforms.

Practical examples

Ignition: Imagine you are diagnosing an ignition system. You capture the waveform of an ignition coil signal and notice an unusual dip. By using annotations, you mark the dip and note its timing relative to other events. Next, you overlay this waveform with a known good ignition coil signal. The overlay makes it immediately apparent that the dip is abnormal, indicating a potential issue with the coil or its connections.

In-cylinder pressure testing: When conducting pressure pulse analysis using an in-cylinder pressure transducer, an event overlay could help you identify the timing of the engine cycles including, intake, compression, ignition/power and exhaust.

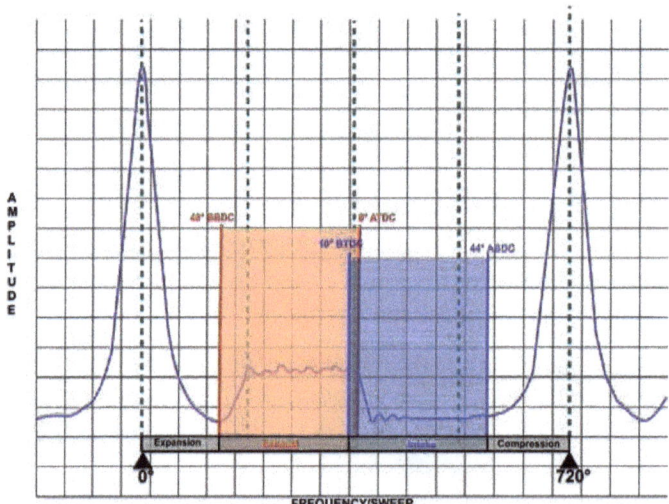

Figure 5.5 In-cylinder pressure screen overlay

Mastering Signal Analysis

Timing Measurements

Accurate timing measurements are crucial for diagnosing **synchronisation** issues in automotive systems. Synchronisation problems can cause a variety of issues, including poor engine performance, increased emissions, and even damage to vehicle components. By using an automotive oscilloscope to perform precise timing analysis, you can effectively troubleshoot and resolve these problems.

Table 5.3 highlights some of the key benefits.

Table 5.3 Timing measurements

Reason	Benefit
Identifying timing discrepancies	One of the primary advantages of precise timing analysis is the ability to identify discrepancies in timing events. For example, if an engine's camshaft and crankshaft are not synchronised correctly, it can lead to misfires, rough idling, and reduced power. By capturing the waveforms of the camshaft and crankshaft sensors, you can compare their timing signals and pinpoint any deviations from the expected synchronisation.
Ensuring proper functioning of engine components	Timing analysis helps ensure that critical engine components are functioning correctly. Components such as ignition coils, fuel injectors, and valves rely on precise timing to operate efficiently. Any deviation in their timing can affect the engine's performance. By analysing the waveforms of these components, you can confirm whether they are operating within the specified time intervals.
Verifying repairs and adjustments	After performing repairs or adjustments, precise timing analysis can be used to verify their effectiveness. You can capture waveforms before and after the repair to ensure that the timing issues have been resolved. This not only provides confidence in the repair but also helps prevent future problems by ensuring that the components are synchronised correctly.
Enhancing diagnostic accuracy	Precise timing analysis enhances the overall accuracy of diagnostics. By having detailed and accurate timing information, you can make more informed decisions about the condition of the vehicle's systems. This reduces guesswork and increases the likelihood of identifying the root cause of synchronisation issues quickly and efficiently. Reference waveforms provide a valuable source of information for this type of activity.
Improving efficiency	Using an oscilloscope for timing analysis can significantly improve the efficiency of the diagnostic process. With the ability to capture and analyse waveforms in real-time, you can quickly identify synchronisation problems and take corrective actions. This reduces frustration, the time spent on diagnostics and increases the productivity of the workshop.

Mastering Signal Analysis

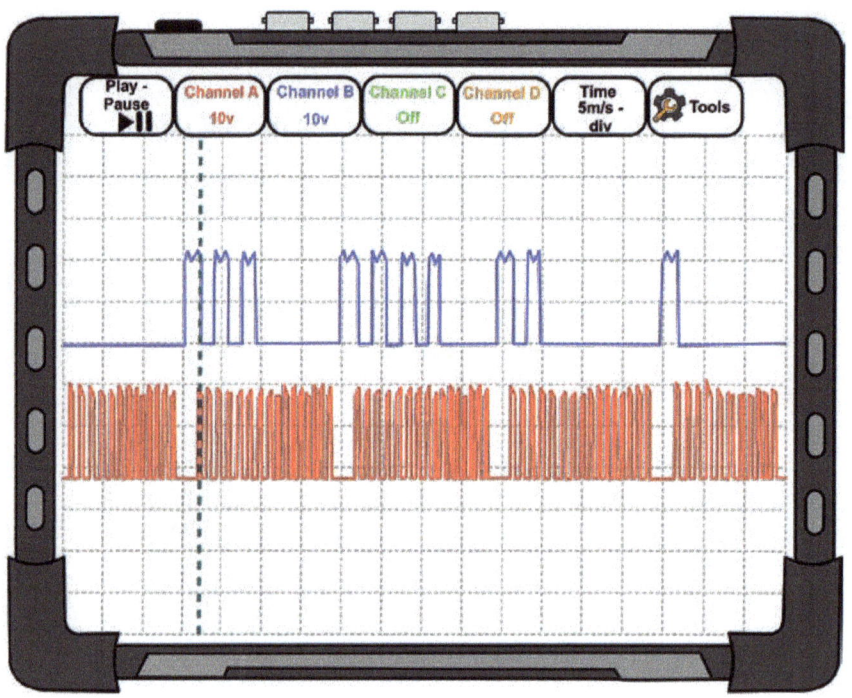

Figure 5.6 Timing measurements

Pulse Width Diagnostics

Pulse Width Modulation (PWM) is a technique used to control the power delivered to various automotive components, such as sensors, actuators, and control modules. Understanding and diagnosing PWM signals is essential for ensuring the proper function of these components, which directly impacts the overall performance and efficiency of the vehicle.

What is pulse width modulation?

Pulse Width Modulation is a method of varying the power to an electrical component by changing the width of the pulses in the signal. The **duty cycle**, which is the proportion of time the signal is active compared to the total period, determines the amount of power delivered. A higher duty cycle means more power is delivered, while a lower duty cycle means less power.

Pulse width modulation PWM often refers to how long something is switched on and is normally represented by a measurement of time.
Although created by pulse width modulation, duty cycle on the other hand, refers to the comparison between the amount of time something is switched on to the amount of time it is switched off (i.e. on-duty is when it's switched on, off-duty is when it's switched off).
Duty cycle is normally represented as a percentage %.

Mastering Signal Analysis

Synchronisation - the precise coordination of various vehicle systems or parts to ensure they operate seamlessly together.

Pulse Width Modulation PWM - a technique that controls the amount of power delivered to vehicle components by varying the width of electronic signal pulses.

Duty cycle - the ratio of active time to total time in a system or component's operation, often expressed as a percentage, to indicate how frequently it is functioning within a given period.

Importance of PWM in automotive systems

PWM is used in various automotive systems to control components such as:

- Fuel injectors
- Ignition coils
- EGR valves
- Turbochargers
- Electric motors
- Lights and indicators

By modulating the width of the pulses, the control modules can precisely regulate the performance of these components, ensuring optimal operation and efficiency.

Measuring PWM signals

Setup

Step 1 • Connect the oscilloscope probes to the component or control module you want to measure.

Step 2 • Set the oscilloscope to the correct voltage and time scale to capture the PWM signal accurately.

Step 3 • Adjust the trigger settings to stabilise the waveform for better analysis.

Step 4 • Once the oscilloscope is set up, start capturing the waveform of the PWM signal.

Step 5 • Observe the waveform carefully, noting the duty cycle and frequency. Some oscilloscopes will allow you to access these figures, calculated directly from the waveform in the measurements menu.

Step 6 • Save the waveform data for further analysis and comparison.

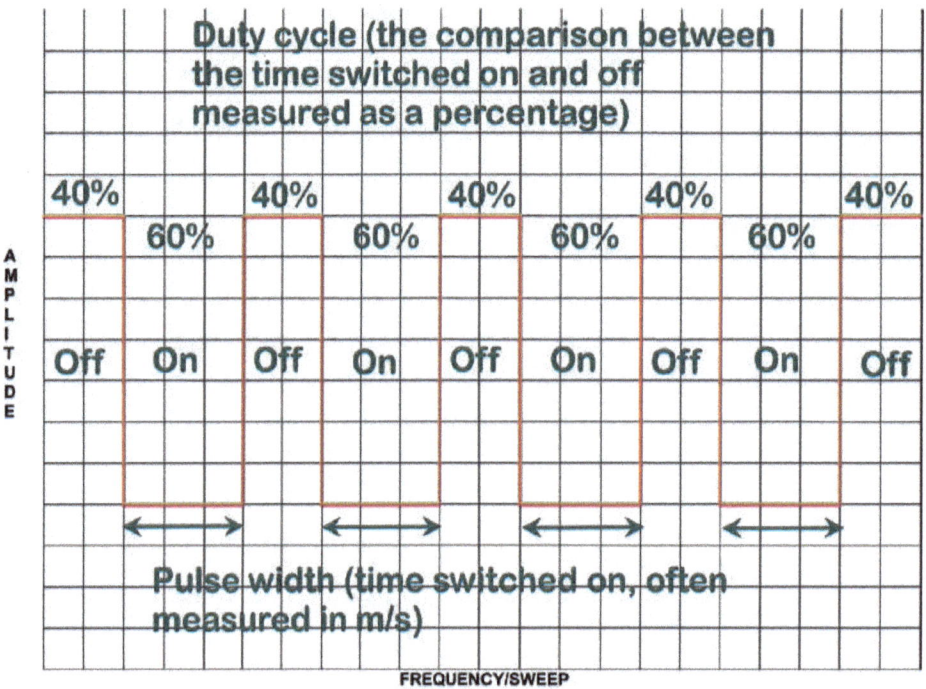

Figure 5.7 Pulse width modulation and duty cycle

Interpreting PWM signals

Understanding the captured waveform is crucial for diagnosing issues with PWM signals. **Table 5.4** shows some key aspects to consider and common issues.

Table 5.4 PWM signals and common issues

Signal	Description	Common issue
Duty cycle	The duty cycle of a PWM signal is the percentage of time the signal is active within one cycle.	An incorrect duty cycle could indicate issues with the control module or the component itself. You should verify that the duty cycle matches the manufacturer's specifications and make necessary adjustments or repairs.
Frequency	The frequency of the PWM signal is the number of cycles per second (normally measured in Hz). It is essential to verify that the frequency is within the expected range. Incorrect frequencies can lead to malfunctioning components and reduced performance.	Frequency variations may occur due to unstable power supplies, malfunctioning control modules, electromagnetic interference (EMI), or inconsistencies in the mechanical system's operation. It is advisable to ensure that the frequency matches the expected specifications and, if discrepancies are found, proceed with further investigation.

Mastering Signal Analysis

Table 5.4 PWM signals and common issues

Signal	Description	Common issue
Amplitude	The amplitude of the PWM signal is the voltage level (height or magnitude) during the active phase of the pulse. Checking the amplitude is vital to ensure that the component receives the correct power level. Variations in amplitude can affect the efficiency and reliability of the component.	Amplitude fluctuations can be caused by voltage drops, poor connections, back electromotive force (EMF) or faulty components. To ensure efficient power delivery, you must check for stable amplitude levels. It is advisable to ensure that the amplitude matches the expected specifications and, if discrepancies are found, proceed with further investigation.
Consistency	Consistency in the PWM signal is critical for the stable operation of automotive components. You must look for any irregularities or fluctuations in the waveform. Inconsistent signals can lead to intermittent issues and unreliable performance.	Distorted PWM signals can be caused by electrical interference, faulty wiring, or malfunctioning control modules. Signal distortion can lead to incorrect power delivery and adversely affect component or system performance.

Using measurement functions to analyse PWM and duty cycle

To accurately assess the performance of PWM signals, you can use oscilloscope measurement functions such as duty cycle analysis, edge count, pulse width, and rise/fall time.

With these measurement functions, you gain deeper insight into PWM system behaviour, allowing precise troubleshooting and component verification. **Table 5.5** shows examples of PWM measurement functions.

Table 5.5 Oscilloscope measurement functions for PWM analysis

Measurement function	Definition	Application in automotive diagnostics
Duty cycle (positive)	Percentage of time the signal is HIGH in one cycle.	Determines actuator response in pull-up circuits.
Duty cycle (negative)	Percentage of time the signal is LOW in one cycle.	Determines actuator response in pull-down circuits.
Edge count (rising)	Number of transitions from LOW to HIGH.	Diagnoses activation patterns in positive switched sensors and actuators.
Edge count (falling)	Number of transitions from HIGH to LOW.	Diagnoses activation patterns in negative switched sensors and actuators.
High pulse width	Duration the signal remains at HIGH state.	Diagnoses OFF states of negative switched actuators, such as idle control or EGR valves.
Low pulse width	Duration the signal remains at LOW state.	Used in solenoid control, fuel injector open time, and motor speed regulation.
Rise time	Time taken for signal transition from LOW to HIGH.	Evaluates switching efficiency in **IGBT** and **MOSFET** driver circuits.
Fall time	Time taken for signal transition from HIGH to LOW.	Helps analyse relay and **transistor** behaviour for smooth switching.

Mastering Signal Analysis

IGBT - Insulated Gate Bipolar Transistor is a high-efficiency semiconductor switch used in electric and hybrid vehicle power electronics.

MOSFET - Metal-Oxide-Semiconductor Field-Effect Transistor is a high-speed switching device used in vehicle power electronics.

Transistor - a semiconductor device used to switch or amplify electrical signals in vehicle systems.

Practical applications of pulse width diagnostics

Pulse width diagnostics have multiple applications in the automotive industry, examples include:

Fuel Injector analysis

You can use PWM diagnostics to measure the duration and timing of fuel injector pulses. This helps ensure that fuel is delivered accurately and efficiently to the engine.

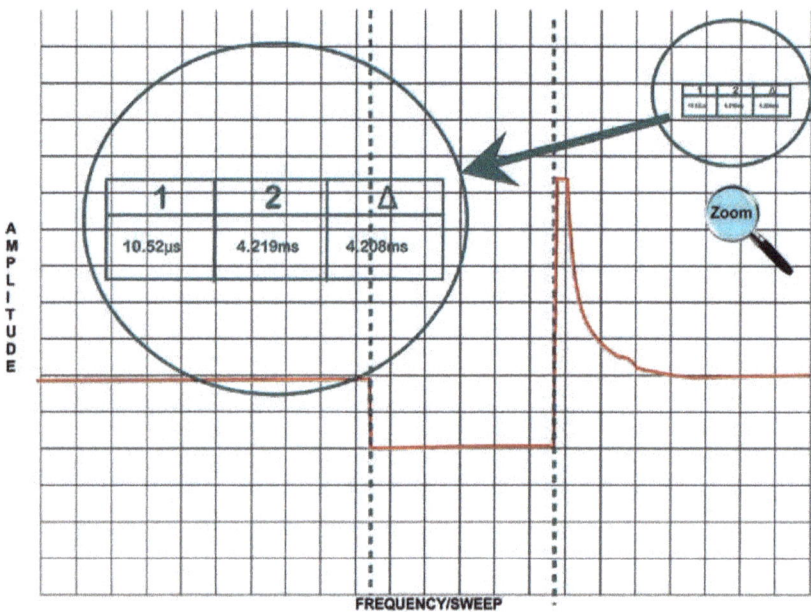

Figure 5.8 Fuel injector waveform pulse width

Remember that the on/off switched pulse used to activate a fuel injector, does not necessarily mean that the injector is mechanically sound. In essence, the circuit could be switching on and off, but the injector pintle may be seized closed for example.
Some fuel injectors will display a pintle hump, in the current ramp or collapsing EMF on a voltage waveform. This can sometimes be used to confirm mechanical operation.

Mastering Signal Analysis

Ignition system testing

By analysing the pulse signals of ignition coils and spark plugs, you can verify the timing and strength of the ignition system. This is crucial for maintaining engine performance and preventing misfires.

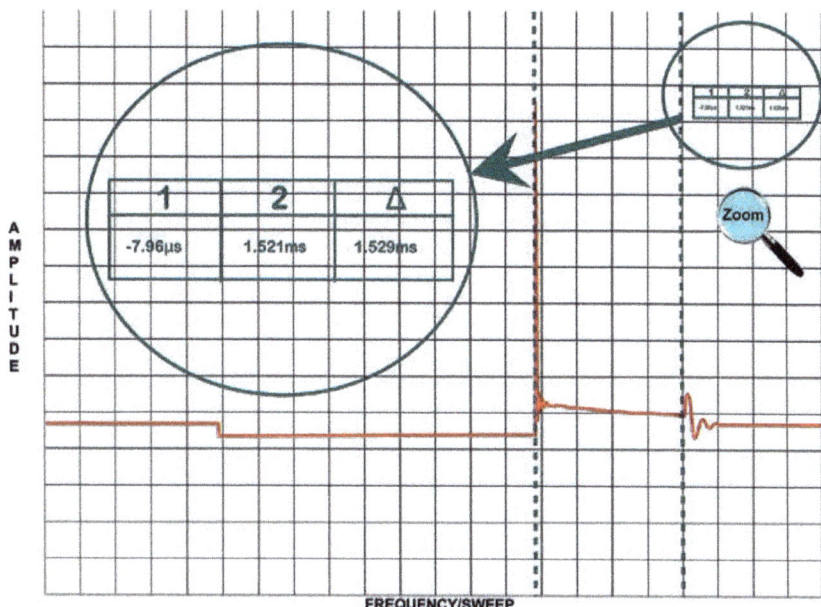

Figure 5.9 Ignition spark duration

Turbocharger control

PWM signals can be used to regulate the operation of turbocharger wastegates, bypass/dump valves and variable vane technology. You can diagnose the PWM signals to ensure optimal turbocharger performance and boost pressure.

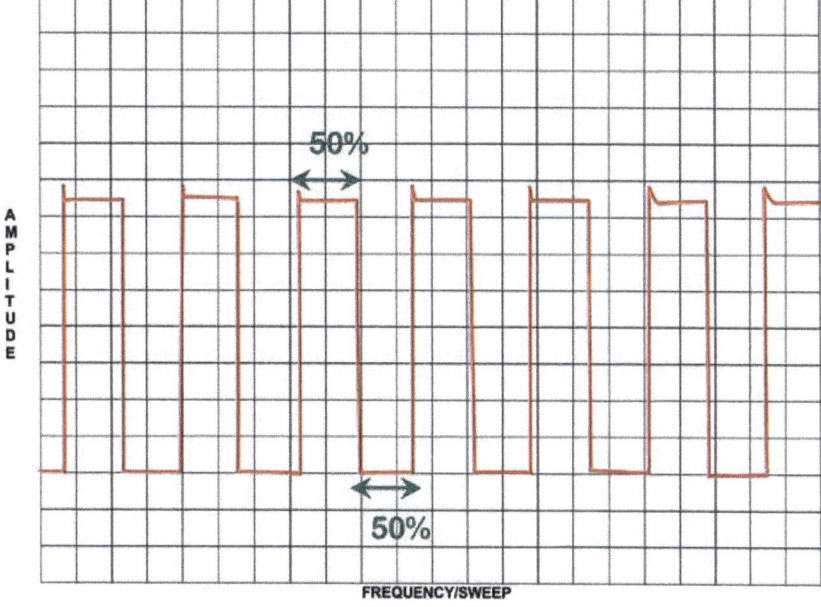

Figure 5.10 Wastegate duty cycle

Electric motor control

PWM diagnostics is essential for analysing the speed and torque control of electric motors. This is particularly important for hybrid and electric vehicles.

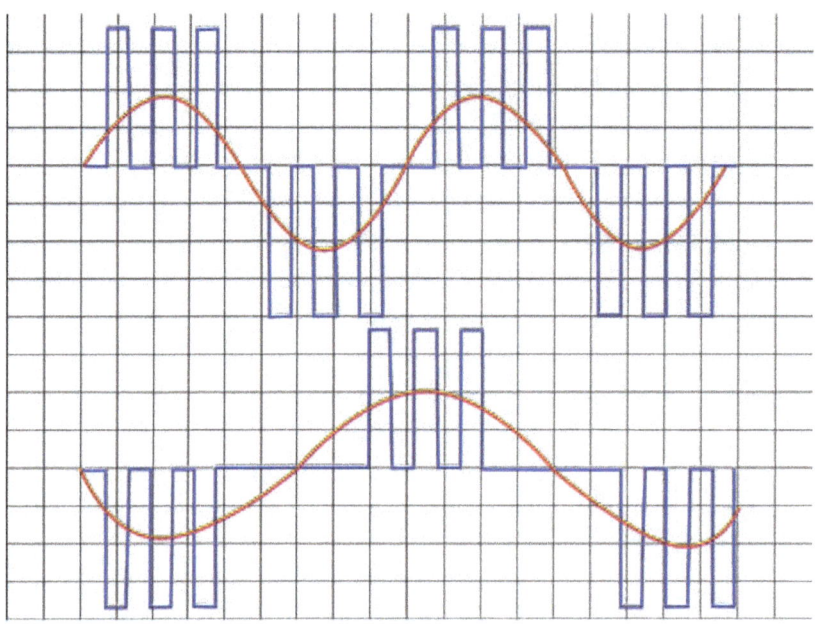

Figure 5.11 Electric motor control

Lighting systems

PWM signals control the brightness and operation of automotive lighting systems. You can use pulse width/duty cycle diagnostics to verify that lights function correctly and efficiently.

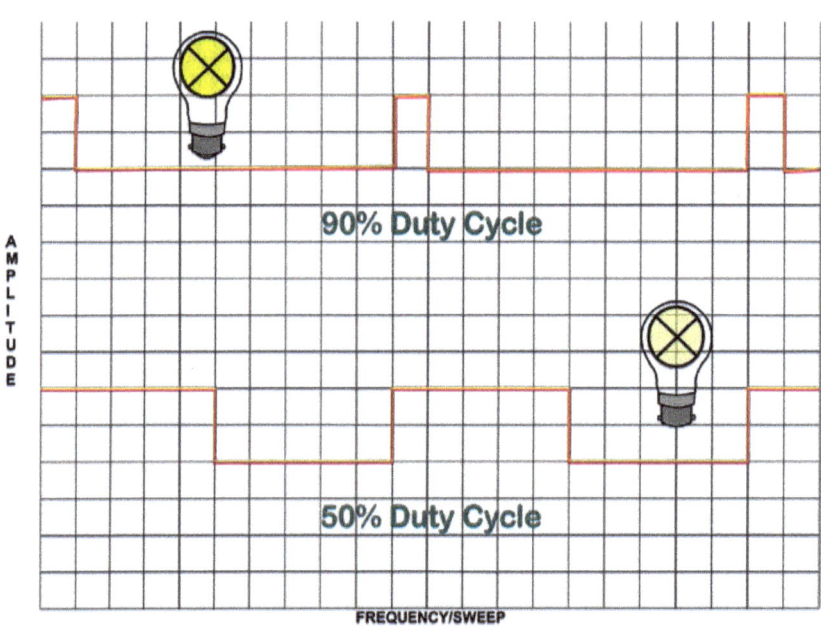

Figure 5.12 PWM lighting control

Mastering Signal Analysis

CAN BUS and In-vehicle Networks

Vehicle operating information is shared between different systems using an in-vehicle **network**.
In a network, the control units, often referred to as **nodes**, are linked by an in-vehicle communication system that allows the transmitting and receiving of data. CAN BUS is probably one of the most widely used networks within vehicle design, and the name 'CAN BUS' has become synonymous with electronic control unit (ECU) communication to the point where it is often used to describe all in-vehicle networking, even if another type is actually being used.

Controller area network (CAN) was introduced by Robert Bosch in the 1980s and is an international standards organisation (ISO) standard for a **serial multiplex** communication **protocol**. CAN BUS is a network communication standard where information is bundled into a '**data packet**' and sent onto the bus system along two twisted wires. Every node on the bus system receives the message and acts if required.

The advantages of CAN BUS are:

- Transmission speeds are much faster than those used in conventional communication (up to 1 **Mbps**), allowing much more data to be sent.
- The system is very immune to interference (noise), and the data obtained from each error detection device is more reliable.
- Each ECU connected via the CAN BUS communicates independently, therefore if an ECU is damaged or faulty, communications can be continued in many cases.

Physical layer of in-vehicle network systems

The **physical layer** is the name given to the wiring of an in-vehicle network system. It is used to join the various nodes or components to each other, and also to connect networks of different speeds and systems. To help locate and trace the network systems, manufacturers create wiring schematics known as **topology diagrams**, showing the layout of the major components.

The CAN BUS line consists of two cables, known as CAN H and CAN L (CAN High and CAN Low, respectively).

The CAN High and Low wires are twisted together, and this helps to cancel out noise which may be caused by electromagnetic interference from other vehicle electrical systems. At the ends of each CAN line are termination resistors that help to dampen out voltage spikes (back EMF) which could be caused as the communication is triggered on and off. The CAN BUS lines connecting two dominant ECUs are the main bus lines, and the CAN BUS lines connecting each individual ECU are the sub-bus lines.
Each ECU communicates with the CAN BUS periodically, sending information from several sensors. This information is circulated on the CAN BUS as a data packet. Each ECU needing data on the CAN BUS can receive these data frames sent from each ECU simultaneously. A single ECU transmits multiple data frames. When data packets conflict with one another (when more than one ECU transmits signals at the same time), data is prioritised for transmission by a process called '**mediation**'.

If mediation is required:

1. The data frame with high priority is transmitted first according to ID codes embedded in the data packet.
2. Transmission of low-priority data is suspended by the issuing ECU until the bus clears (when no transmission data exists on the CAN BUS).
3. The ECU containing suspended data frames transmits when the bus becomes available.

Mastering Signal Analysis

Network - the connection and coordination of various vehicle systems or parts to ensure they operate seamlessly together.

Nodes - an electronic control unit (ECU) or sensor within a vehicle's network that communicates via protocols like CAN, LIN, or Ethernet to exchange data and manage functions such as powertrain control, diagnostics, and infotainment.

Controller area network - a robust vehicle communication protocol that enables ECUs and sensors to exchange real-time data without a central computer.

Serial - a communication method using sequential (one after another in sequence) data transmission.

Multiplex - a communication system that allows multiple electronic control units (ECUs) to share data over a single network.

Protocol - a standardised communication method that enables data exchange between electronic control units (ECUs), sensors, and actuators in a vehicle.

Data packet - a structured unit of information transmitted between electronic control units (ECUs) and sensors in a vehicle, carrying essential data like commands, sensor readings, or diagnostics.

Mbps - megabits per second is a measure of data transfer speed in vehicle networks, indicating how many million bits of information are transmitted per second in communication protocols.

Physical layer - the hardware and electrical signalling framework of a vehicle communication network.

Topology diagrams - a visual representation of a vehicle's electronic network, illustrating how electronic control units (ECUs), sensors, and communication links are interconnected.

Mediation - the process of facilitating communication or resolving conflicts between electronic control units (ECUs), networks, or vehicle systems.

Think of CAN BUS like a group chat for vehicle components.
- Each participant (ECU or sensor) can send messages, like reporting speed, fuel levels, or airbag status.
- Instead of every person having to text each other individually, CAN BUS acts as a shared group chat, where messages are sent once and everyone listens in.
- The chat moderator (arbitration/mediation process) ensures that urgent messages—like braking commands—get posted first, while less critical updates—like window position—wait their turn.
- Unlike older systems where components needed dedicated wires for communication (like texting separately), CAN BUS reduces wiring complexity by letting everyone use the same network, making vehicle communication faster, more efficient, and less cluttered.

Mastering Signal Analysis

Communication data of in-vehicle network system

When an ECU receives a signal from a vehicle sensor, it processes it and places the information on the network bus as a data packet. The data packet is usually made up of the following components:

- A header, **SOF (Start of Frame)**: the equivalent of 'hello, I am transmitting a message'.
- The priority **ID (Identifier)** region: how important this message is.
- Remote transmission request **RTR**: indicates whether a message is a data frame or a remote request frame. A data frame contains actual data, while a remote request frame asks for data from another ECU.
- Flexible data frame **FDF**: indicates whether a message is a CAN FD frame or a Classical CAN frame. CAN FD is an extension of the CAN protocol that allows for higher data rates and longer data fields.
- Data length **Control region DLC**: this tells the receiver how many bytes of data are in the packet.
- Data type **Data region**: this indicates what type of information is contained, e.g. voltage, speed, temperature, etc.
- Data **Data region**: the actual sensor information itself.
- An error detection code **Cyclic Redundancy Check (CRC)** region: this verifies that all the information has been received correctly.
- End of message **EOF (End of Frame)**: 'goodbye'.
- Finally, a request for a response from the receiving ECU **ACK (Acknowledge) region**: this says, 'thank you, I got your message'.

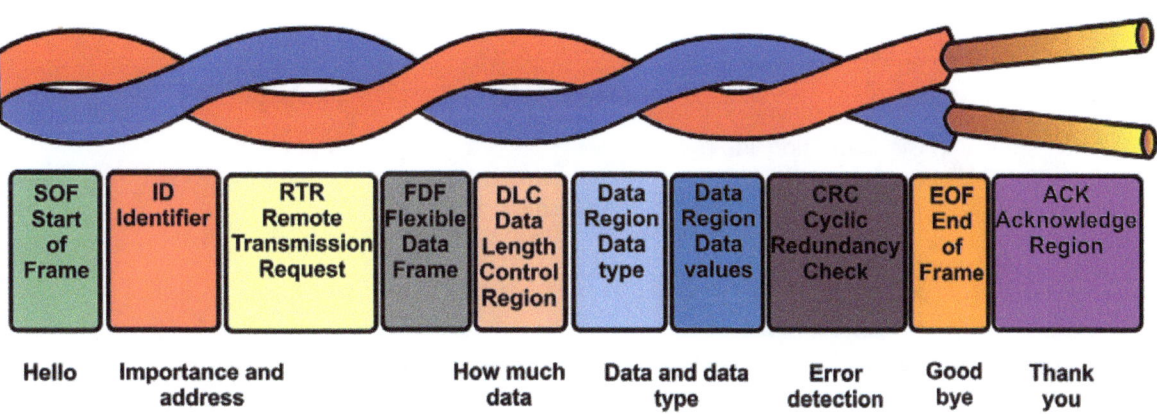

Figure 5.13 CAN BUS communication data packet

Reducing data corruption in CAN BUS systems

To help reduce the possibility of data corruption caused by misinterpretation or external electromagnetic interference, a CAN BUS system uses two communication wires instead of one, twisted over and over each other in a spiral.

The same data is sent on both of these communication wires as an on and off voltage signal.
One signal is sent as a positive switch and one is sent as a negative switch, providing a mirror image on each network wire, which are known as CAN High and CAN Low.
The potential difference between the voltages on the two lines produces a digital signal that can be processed into information.

Data transmission - high speed

The transmitting ECU sends switched voltage through the CAN H and CAN L bus.

It sends 2.5 to 3.5 volt signals to the CAN High line and 2.5 to 1.5 volt signals to the CAN Low line.

The receiving ECU reads the data from the CAN lines as a potential difference of between 3.5 and 1.5 volts.

In **Figure 5.14**, 'Recessive' refers to the state where both CAN H and CAN L are at the 2.5 volt state, and 'Dominant' refers to the state where CAN H is at the 3.5 volt state and CAN L is at the 1.5 volt state. These values correspond to a binary value of either 1 or 0.

Recessive = Logic value of 1
Dominant = Logic value of 0

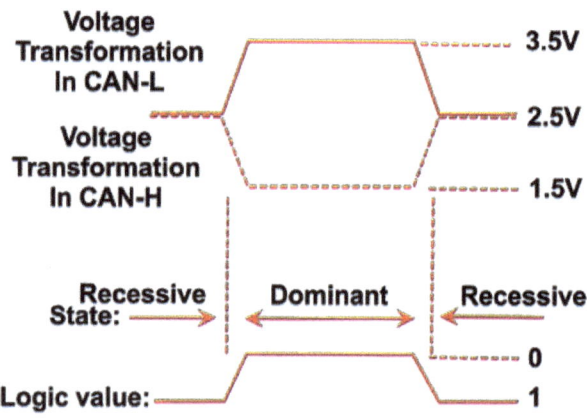

Figure 5.14 Data Transmission High Speed

Data transmission - low speed

The transmitting ECU sends switched voltage through the CAN H and CAN L bus.

It sends 0 to 4 volt signals to the CAN High line and 1 to 5 volt signals to the CAN Low line.

The receiving ECU reads the data from the CAN High and CAN Low as a potential difference of between 5 and 0 volts.

In **Figure 5.15**, 'Recessive' refers to the state where CAN H is at 0 volt and CAN L is at 5 volt, and 'Dominant' refers to the state where CAN H is at 4 volt and CAN L is at 1 volt.

Recessive = Logic value of 1
Dominant = Logic value of 0

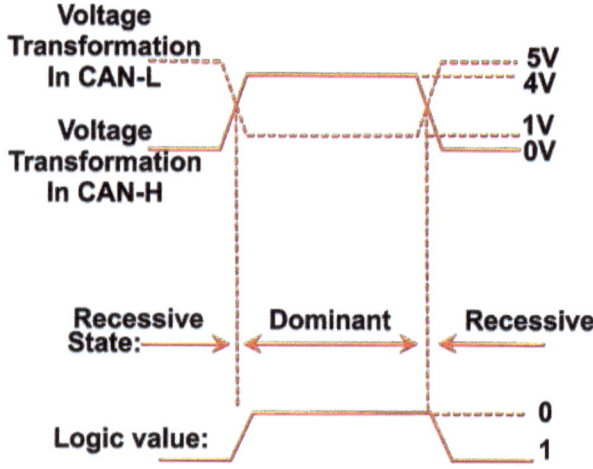

Figure 5.15 Data Transmission Low Speed

Acquiring CAN BUS signals from the vehicle data link connector (DLC)

In order to capture CAN BUS signals from the vehicle data link connector you need to understand the pin layout of an E-OBD (OBD-II) configuration. Pin 6 is connected to CAN High and Pin 14 is connected to CAN Low; these pins carry differential CAN BUS data at 500 kbps (typical for high-speed CAN).

Mastering Signal Analysis

It is often possible to conduct an initial oscilloscope diagnosis of network systems at the pins of the vehicle data link connector. Due to the standardised layout of the 16-pin connector the terminals can be identified from the image shown below:

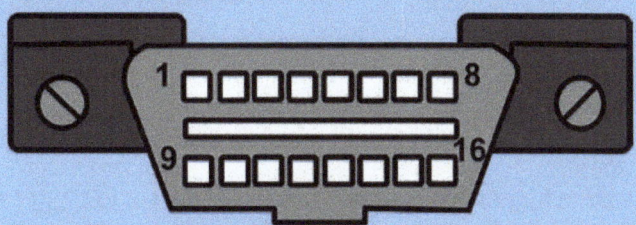

1. Manufacturer specific [sometimes used for network communication].
2. Bus positive SAE J1850 PWM and VPW.
3. Manufacturer specific [sometimes used for network communication].
4. Chassis ground.
5. Signal ground.
6. CAN High.
7. K-Line of ISO9141-2 and ISO14230-4.
8. Manufacturer specific [sometimes used for network communication].
9. Manufacturer specific [sometimes used for network communication].
10. Bus negative SAE J1850 PWM.
11. Manufacturer specific [sometimes used for network communication].
12. Manufacturer specific [sometimes used for network communication].
13. Manufacturer specific [sometimes used for network communication].
14. CAN Low.
15. L-Line of ISO9141-2 and ISO14230-4.
16. Battery voltage.

Always use non-invasive probing methods or a breakout box to avoid damaging terminals. You may need to 'wake up' the BUS by turning the ignition ON or opening a door if needed. Use CAN decoders if your scope supports it, to see actual message IDs and data. Verify termination resistors (\~60 ohms across pins 6 & 14) when the BUS is powered off.

Termination resistances can give a good indication of correct circuit operation. If an ohmmeter is connected in parallel across CAN High and CAN Low (using pins 6 and 14 of the data link connector for example) with the circuit power switched off, then the total recorded resistance will be halved. With 120Ω termination resistors:

- If 60Ω is shown, CAN High and CAN Low should be OK.
- If O/L (infinity) is shown, an open circuit exists in both lines.
- If 0Ω is shown, a dead short exists.
- If 120Ω is shown, one CAN line may be at fault (confirm communication using an oscilloscope).

Mastering Signal Analysis

Setup and use

Step 1
- Connect the Scope:
- Channel A: Probe connected to pin 6 (CAN High).
- Channel B: Probe connected to pin 14 (CAN Low).
- Ground: Connect to pin 4 or pin 5 (chassis or signal ground).

Step 2
- Configure Oscilloscope Settings:
- Time base: Start at 20 µs/div to 50 µs/div.
- Voltage range: Channel A: 0 to 5V. Channel B: 0 to 5V (if preferred, invert Channel B for easier differential viewing).
- Trigger: Use edge triggering on Channel A (CAN H), set to trigger at \~3V.
- Coupling: DC. - Sample rate: At least 1 MS/s.

Step 3
- Capture the CAN Bus Waveform:
- Once configured, you'll see both:
- CAN High (CH A): idles at \~2.5V, rises to \~3.5V (recessive to dominant).
- CAN Low (CH B): idles at \~2.5V, drops to \~1.5V.

Step 4
- Analyse the Waveform: Look for:
- Normal CAN Communication.
- Clear, clean square-wave signals on both lines.
- Differential voltage of \~2V during dominant state.
- Symmetrical timing between High and Low.

Step 5
- Analyse the Waveform: Fault Indicators:
- Flat-line at 2.5V: no communication (bus off or module asleep).
- Only one line changing: wiring or module fault.
- Over/under voltage: short to power or ground.
- Reflections, ringing: termination resistor issues.

Step 6
- Compare Against Known Good: If available, load a reference waveform for your specific vehicle make and model to identify any anomalies.

Step 7
- Save and Document Findings:
- Save the waveform capture.
- Annotate and timestamp observed issues or good performance.
- Note vehicle info, key-on status, and any connected modules.

Network diagnosis of in-vehicle network systems

If a critical network failure occurs, such as a short to positive or ground, the vehicle may suffer a complete communication loss. With a networked system, if communication is lost within a certain area, a number of items will not work, and numerous trouble codes may be generated.

Mastering Signal Analysis

Having connected a scan tool and retrieved the diagnostic trouble codes, you should look for the code that is the root cause. Communication failures are normally an effect of the original fault (i.e. 'unable to communicate' or 'communication lost'). You should ask yourself, 'Is this the cause or an effect created by the fault?'

CAN BUS systems report communication faults as live data. As a result, once you have identified the causal trouble code, you may be able to conduct a diagnosis by disconnecting and isolating components or sections of the low-voltage wiring loom until communication is re-established. With the oscilloscope connected and running watch the waveform communication as the network components are isolated.

Oscilloscope testing of CAN BUS

A CAN system can often be identified as a pair of twisted wires entering or leaving an ECU. An oscilloscope can be connected to these wires by 'back probing' at the ECU socket, but if the vehicle is operating using CAN BUS, a good place to connect to the main circuit is at pins 6 and 14 of the diagnostic socket (DLC).

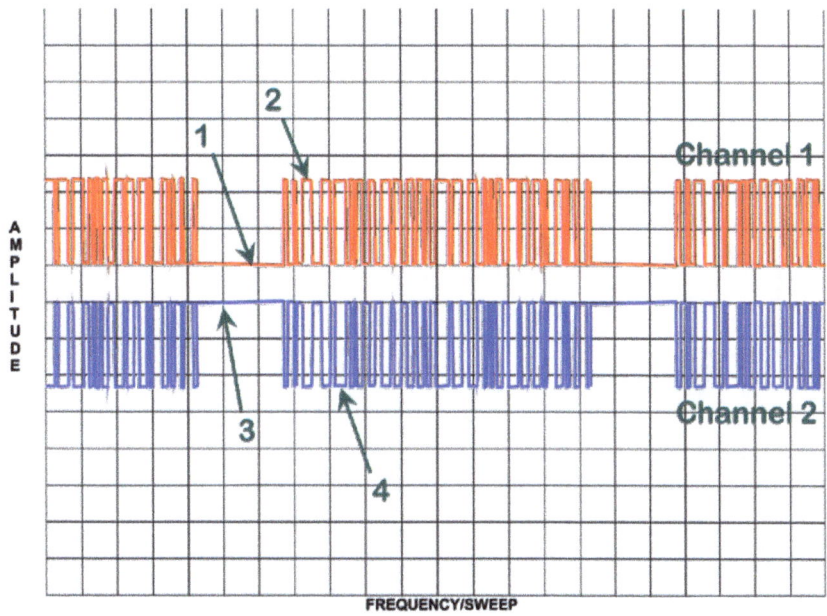

Figure 5.16 CAN BUS (CAN H and CAN L)

Table 5.6 Waveform analysis CAN High and CAN Low

Waveform component	Description
1	**Channel 1** is connected to CAN H (High) and switches positively. This means that the voltage is 0 or 2.5 volts in the off position depending on network speed.
2	**Channel 1**. When switched on, the voltage will jump to 3.5 or 4 volts depending on network speed.
3	**Channel 2** is connected to CAN L (Low) and switches negatively. This results in a voltage of 5 or 2.5 volts in the off position depending on network speed.
4	**Channel 2**. When switched on, the voltage will fall to 1 or 1.5 volts depending on network speed.

Mastering Signal Analysis

By changing the frequency/sweep on the oscilloscope and aligning the voltage amplitudes between Channel 1 and Channel 2, it is possible to compare the two patterns and see the potential difference from CAN High and CAN Low. This can then be interpreted as a dominant or recessive logic value. *(See Figures 5.14 and 5.15).*
Recessive = Logic value of 1
Dominant = Logic value of 0

It is important to check that the patterns from CAN High and Low show equal and opposite with clean edges when examining the waveform. This indicates that the network wiring circuit is operating effectively, and that any non-responsive individual ECU is likely caused by the ECU itself.

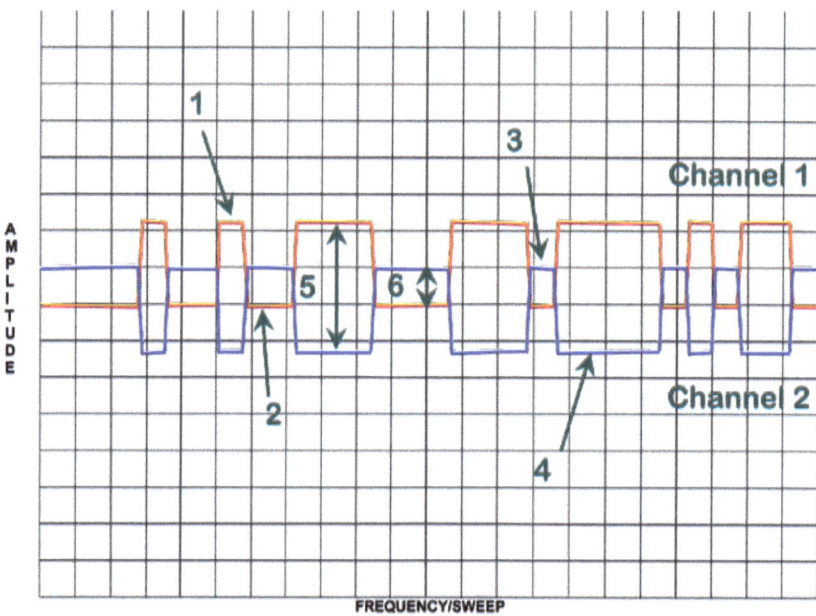

Figure 5.17 CAN BUS potential difference

Table 5.7 Waveform analysis CAN BUS potential difference

Waveform component	Description
1	**Channel 1** is connected to CAN H (High) and switches positively. This means that the voltage is 3.5 or 4 volts in the on position depending on network speed.
2	**Channel 1.** When switched off, the voltage will fall to 0 or 2.5 volts depending on network speed.
3	**Channel 2** is connected to CAN L (Low) and switches negatively. This results in a voltage of 5 or 2.5 volts in the off position depending on network speed.
4	**Channel 2.** When switched on, the voltage will fall to 1 or 1.5 volts depending on network speed.
5	This section of the waveform shows a dominant logic value of 0.
6	This section of the waveform shows a recessive logic value of 1.

Mastering Signal Analysis

Calculating network speed using an oscilloscope

It is sometimes necessary to know the speed of the network that you are diagnosing. This is especially useful when setting up a serial decoder *(see Chapter 6)*.

Setup and Use

Step 1
- Connecting Probes: Attach the oscilloscope probes to the CAN High and CAN Low lines. These connections allow the oscilloscope to capture the differential signals transmitted across the network.

Step 2
- Start Communication: Switch on the vehicle/network and adjust time base and amplitude to acquire a stable capture. Pre-sets or guided tests can often help with this step and then be fine-tuned for accurate signal acquisition.

Step 3
- Analyse the waveform: Pause the capture and use the magnifier tool to zoom in on a section which shows a single data bit. This can often be identified as the narrowest dominent bit as illustrated in **Figure 5.18**.

Step 4
- Measure the data bit: Place rulers at the start and end of this single data bit and measure the time difference Δ (Delta). At this point, some oscilloscope software may calculate the network speed automatically.

Step 5
- To calculate the network speed use the following equation: One divided by the delta time, and then multiplied by one thousand to show the result in kilohertz.
- $1/Δ × 1000 = kHz$

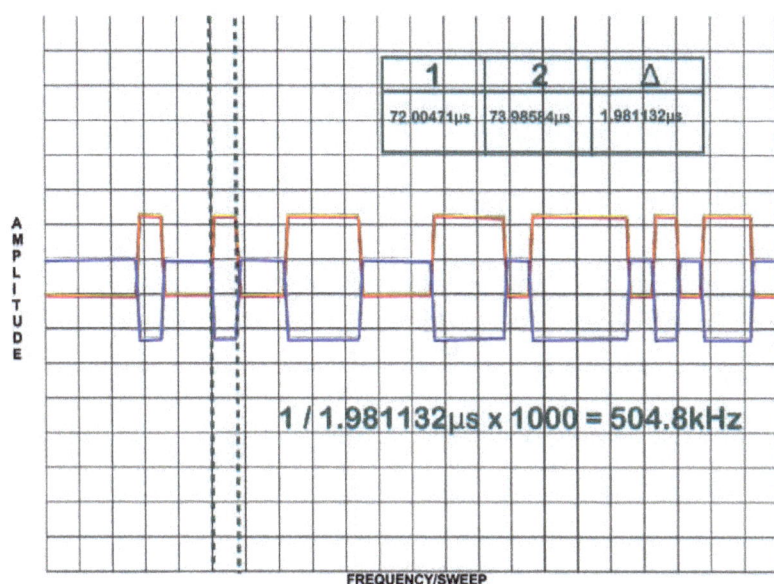

Figure 5.18 Using an oscilloscope to calculate network speed

Mastering Signal Analysis

SAFETY: Using insulation piercing probes to measure CAN BUS signals is not recommended, because this can damage the integrity of the wiring and promote communication problems.

CAN BUS edge matching:

To prevent data corruption, CAN High and CAN Low must be synchronized and aligned. This can be evaluated using a method called edge matching.
Capture a CAN BUS trace from both High and Low channels.
Pause the waveform and use a ruler to verify the edge alignment.

Table 5.8 shows examples of some CAN BUS issues.

Table 5.8 CAN BUS faults and their effects on oscilloscope waveforms

Fault type	Waveform effect on oscilloscope	Diagnostic implications	Waveform example
CAN H to CAN L (direct short between bus lines)	The differential signal collapses to zero volts, eliminating proper communication. The scope shows a flat-line or erratic signal.	Total network failure: modules lose communication.	
CAN H to CAN L (cross connection)	The signals are identical rather than being mirror images.	There is a fault, but it might not cause operational errors.	

Mastering Signal Analysis

Table 5.8 CAN BUS faults and their effects on oscilloscope waveforms

Fault type	Waveform effect on oscilloscope	Diagnostic implications	Waveform example
CAN H to ground	CAN H voltage drops, distorting the differential signal, causing erratic voltage swings or a weak signal.	Poor network performance, intermittent communication errors.	
CAN L to ground	CAN L voltage collapses, reducing the differential voltage and causing decoding errors.	Possible loss of lower-priority messages, erratic ECU behaviour.	
CAN H to power (+12V or battery)	Excess voltage pushes CAN H beyond acceptable levels, creating overvoltage trace.	Can damage transceivers, cause BUS shutdown.	
CAN L to power (+12V or battery)	CAN L voltage increases, creating an abnormally high differential voltage, leading to decoding errors.	Erratic data transmission, possible hardware damage.	

Table 5.8 CAN BUS faults and their effects on oscilloscope waveforms

Fault type	Waveform effect on oscilloscope	Diagnostic implications	Waveform example
Open circuit (CAN H or CAN L disconnected)	The affected signal flatlines, showing unstable or missing waveforms. If both lines open, the differential voltage may be near zero, causing BUS errors.	Communication loss or severely degraded signal integrity.	

 Please note that the examples provided in **Table 5.8** may not encompass all potential CAN BUS faults or issues. Given the multitude of possible errors and faults, any deviation from an established good example should be thoroughly investigated to determine the root cause of the issue.

Application-Specific Diagnostics

Application-specific diagnostics involves focusing on particular systems within the vehicle and tailoring diagnostic techniques to meet the specific requirements of these components. For example, assessing the performance of hybrid and electric vehicles (EVs) requires specialised knowledge and tools to test critical components such as inverters, battery management systems, and motor controllers. The complexities of these systems necessitate a thorough understanding of waveform analysis to identify potential issues accurately. Similarly, other systems require distinct diagnostic methods, each designed to detect and address specific problems.

Hybrid and EV system diagnostics

Working on hybrid and electric vehicles (EVs) presents unique challenges and hazards. You must exercise caution and adhere strictly to safety protocols when working on or around the high-voltage systems to prevent serious injury or death.

Work and diagnostic testing of high-voltage systems should only be conducted if you have received adequate training, using the correct Personal Protective Equipment (PPE), fully insulated tooling, and correctly rated and calibrated diagnostic electrical test equipment.

The following diagnostic descriptions are designed to support knowledge and understanding, but do not act as a substitute for appropriate training. Never attempt any diagnosis or repairs unless you are suitably qualified and have the correct tools, equipment, and safety measures in place. When conducting diagnosis on 'live' high-voltage systems, try to avoid lone working.

Mastering Signal Analysis

Setup and testing of components

The successful diagnosis of hybrid and EV systems requires a thorough understanding of the components involved and the use of specialised diagnostic equipment. The following sections describe the function of some components found in hybrid and electric vehicles and gives examples of tests that could be conducted using an oscilloscope.

 Due to the dangers involved with testing high voltage systems, test procedures will be restricted to a basic overview. In order to diagnose faults, you should undertake specific EV training and have access to certified equipment and Personal Protective Equipment. Further details can be found in the textbook *Principles of Electric Vehicle Technology*.

Contactors and relays

Contactors and relays are electrically controlled switches used for power distribution within EVs. They control many of the high voltage systems, including those used for connecting and disconnecting the drive battery during startup and shutdown. These are known as system main relays (SMR).

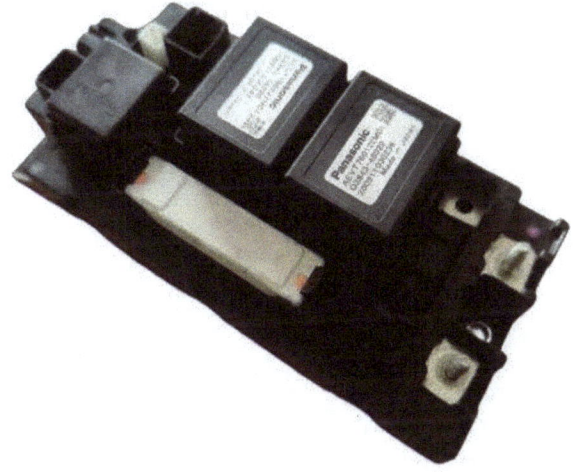

Figure 5.19 System main relays SMR

 Testing contactors or system main relays will often require the battery or high-voltage system covers to be removed. These covers are designed to prevent accidental contact with exposed high-voltage connections. Sometimes, high-voltage interlock loops are used as additional protection, and these may require bypassing in order for any testing to take place. Be aware that bypassing any safety system means that exposed live electrical connections can easily cause injury or death.

Mastering Signal Analysis

Setup and use

Step 1
- Wearing correct PPE, an oscilloscope can be connected to the low-voltage SMR control circuits to measure voltage switching at the contactor terminals. This can often be accessed externally to the main drive battery casing. Follow manufacturers instructions and procedures.

Step 2
- Evsure that the oscilloscope and probes are in good condition and correctly rated for the system being tested. Activate the contactors and observe the waveform to ensure proper operation.

Step 3
- The waveform should display the switching on and off of the pre-charge circuits, and the main power contactors. The results will vary between manufacturers, so the waveform should be referenced to a 'known good'.

Step 4
- The switching can be used to confirm SMR diagnostic trouble codes (DTC).

The following image shows the oscilloscope readings for each channel during the start-up, ready mode, and shutdown sequences of a positive pre-charge system circuit. The image also shows the points on the waveform where the contactors or SMRs are switched on or off, and the checks for possible welding.

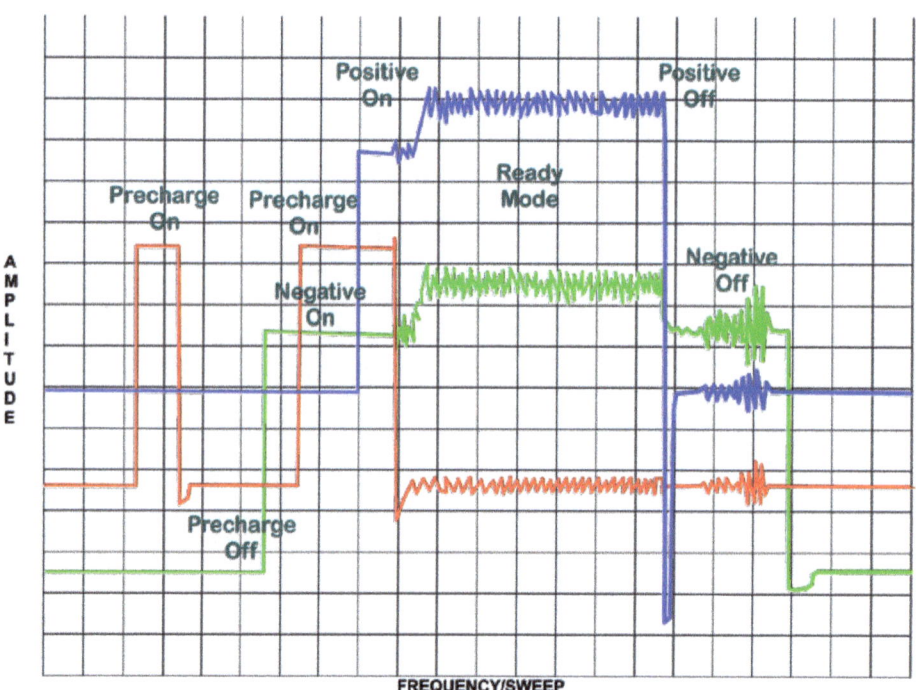

Figure 5.20 Positive pre-charge circuit waveform

Mastering Signal Analysis

To verify the results of diagnostic trouble codes (DTCs) related to system main relays and contactors using an oscilloscope, it is necessary to connect the oscilloscope to the low-voltage circuit.

If a fault exists in the system main relay start-up circuits, the vehicle will not power up or enter ready mode and a malfunction indicator light (MIL) should be illuminated on the driver's display. To allow the start-up sequence to run, you will need to clear any DTCs from the system memory. When you press the start button, the start-up sequence will begin. If a fault is detected, the waveforms on the oscilloscope will flatline at the point of failure, allowing you to confirm the DTC.
If a fault exists in the system main relay shutdown circuits, the vehicle will power down but should finish with a MIL illuminated on the driver's display the next time the vehicle is powered up.
 When you press the start button to switch the vehicle off, if a fault is detected, the waveforms on the oscilloscope will flatline at the point of failure, allowing you to confirm the DTC.

You will only get one opportunity between clearing codes to record and analyse waveforms.

Resolvers

Resolvers are sensors used to determine the position and speed of the drive motor shaft and provide feedback for motor control.

Its construction consists of two main components, a stator, and a rotor.
The stator consists of coils mounted on a frame aligned with the motor generator stator unit. One coil generates a continuous sine wave from an electronic control unit, while two sensor pick-up coils, wound in opposite directions, are positioned at 90-degree intervals. The rotor, fixed to the motor/generator rotor, alters the air gap between exciter and sensor coils as it turns within the resolver stator section. The stable continuous **sinewave** produced by the **exciter** has its electromagnetic signature transferred to the pickup sensor coils via the shaped metal rotor through **variable reluctance**, resulting in two output signals, one with a sinewave signature and one with a cosine signature.

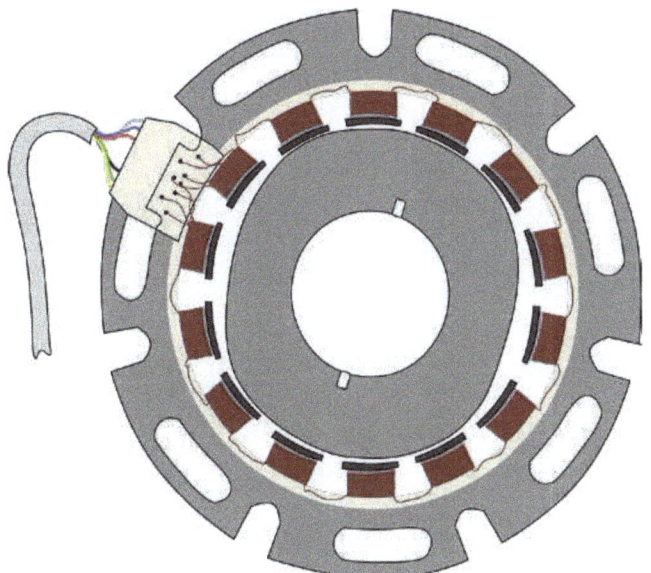

Figure 5.21 A resolver

Mastering Signal Analysis

Setup and use

Step 1
- Wearing correct PPE, prepare to connect the oscilloscope to the resolver input and output signals. Resolvers are highly sensitive, so where possible use a good quality breakout and **floating ground**.

Step 2
- Access to manufacturer information will be required to help identify the resolver signal circuits and exciter.

Step 3
- Connect three oscilloscope channels using a floating ground across the exciter coil, and both signal receiver coils.

Step 4
- Observing all safety precautions, the vehicle will need to be placed in **ready mode** and the motor generator rotated; observe the waveform.

Step 5
- Look for consistent exciter, sine and cosine signals that indicate accurate position feedback. The results will vary between manufacturers, so the waveform should be referenced to a 'known good'.

Although this test is conducted on a low-voltage system at the resolver, precautions should be taken when working on or around high-voltage electrical systems, including the correct use of high-voltage Personal Protective Equipment (PPE).

When testing a resolver, a **differential measurement** needs to be taken across each of the inductive coils. This means that oscilloscopes using a common ground are unsuitable for this form of test. Only oscilloscopes with a floating ground can be used, or those connected with a differential probe.

Depending on the configuration of the vehicle being tested, hybrid or fully electric, the engine may have to be operated while stationary or the vehicle may need to be appropriately raised and safely supported so that the driving wheels can be rotated by hand. Always take appropriate precautions depending on which type of test is being conducted.

Sinewave - a smooth, periodic oscillation that is mathematically described by the sine function.

Exciter - a device, often a small generator or a battery, that supplies the electric current used to produce the magnetic field in another generator or motor.

Mastering Signal Analysis

Variable reluctance - a sensing principle where a magnetic field changes as a magnetic target moves, generating a variable voltage.

Floating ground - a reference point for electrical potential in a circuit that is galvanically isolated from the actual earth ground.

Ready mode - a vehicle state where an electric vehicle (EV) is powered on and prepared for driving.

Differential measurement - the process of measuring the difference between two points, pressures, or electrical potentials.

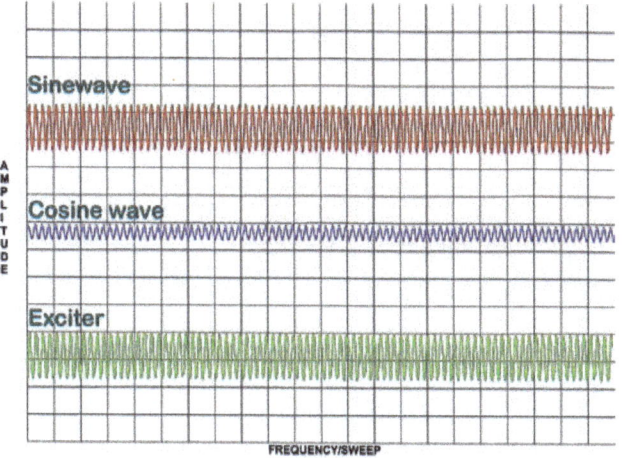

Figure 5.22 Resolver stationary waveform

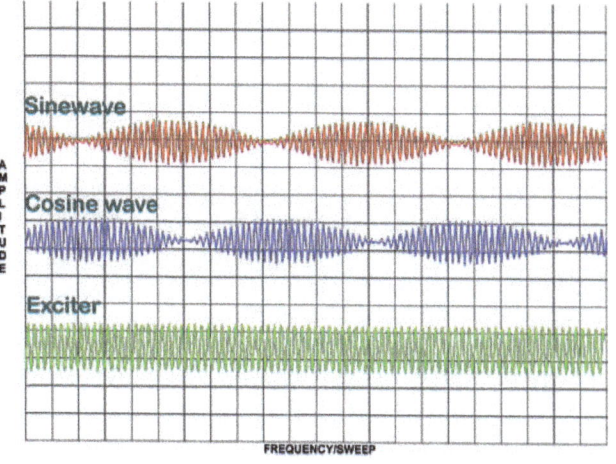

Figure 5.23 Resolver turning waveform

EV charging systems

EV charging systems include onboard chargers (OBC) and external charging stations known as electric vehicle supply equipment (EVSE). A limited amount of oscilloscope testing and diagnosis can be conducted on the charging operation between the EVSE and vehicle, however, this will require the use of a dedicated electric vehicle supply equipment breakout box.

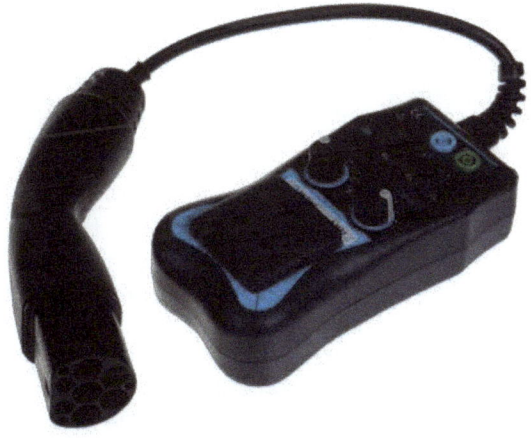

Figure 5.24 EVSE breakout box

Mastering Signal Analysis

Testing these systems involves:

- Measuring the voltage and current delivery during the charging process.
- This can normally be done measuring the signal at the charging system control pilot (CP).
- Using the oscilloscope to capture charging cycles and identify any anomalies.
- Ensuring that the charger operates within the specified parameters.

With the oscilloscope connected via the breakout-box and running, connect the charging cable from the EVSE to the vehicle, capture and analyse a waveform.

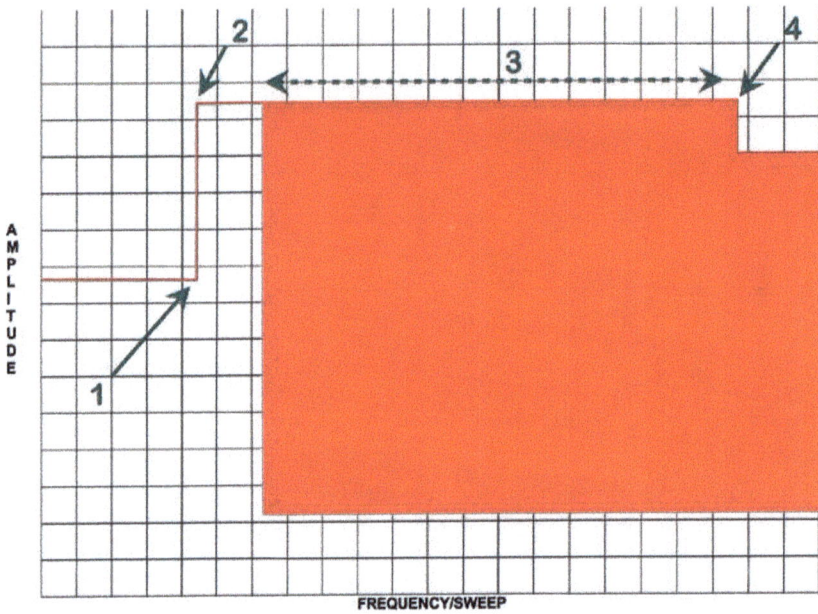

Figure 5.24 CP connected communication waveform

Table 5.9 CP connected communication

Waveform component	Description
1	This is the point on the waveform where the EVSE charge cable is disconnected and the voltage on the control pilot (CP) line is at 0 volts.
2	This is the point on the waveform where the charging cable is connected and the voltage on the control pilot (CP) line jumps to approximately 9 volts.
3	This section shows the start of bidirectional communication between the vehicle and the EVSE, lasting for about 6 seconds. This communication uses a pulse-width modulation with a frequency of around 1,000 Hz and an amplitude modulation of between +9 volts and -12 volts. (*The communication duty cycle is too fast to view without zooming in, see Table 5.10*).
4	This point on the waveform shows where charging has started, and the voltage drops to a modulated amplitude between +6 volts and -12 volts. Bidirectional communication continues and the duty cycle of the pulse width varies with the current delivered. (*The communication duty cycle is too fast to view without zooming in, see Table 5.10*).

Mastering Signal Analysis

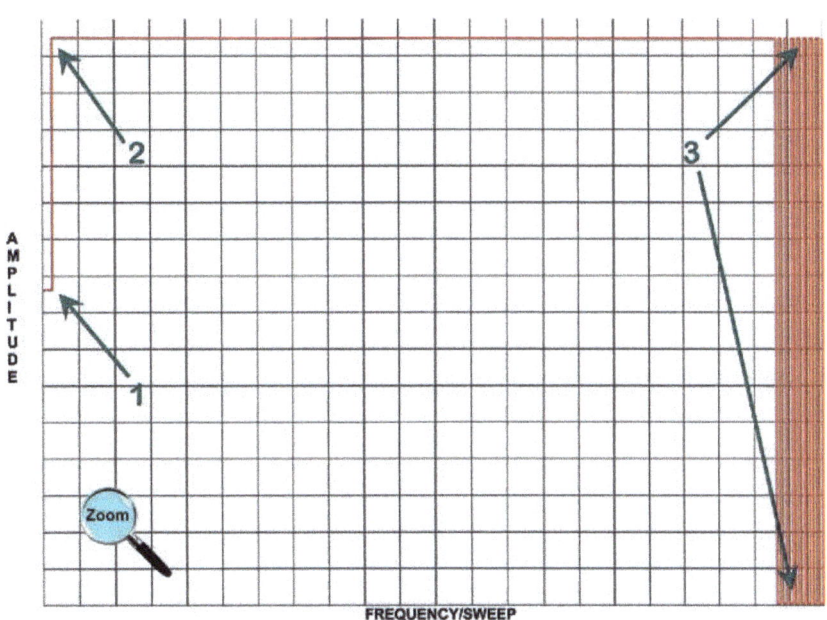

Figure 5.25 CP communication start (zoomed)

Table 5.10 CP connected communication Start (zoomed)

Waveform component	Description
 1	This is the point on the waveform where the EVSE charge cable is disconnected and the voltage on the control pilot (CP) line is at 0 volts. (Zoomed in).
 2	This is the point on the waveform where the charging cable is connected and the voltage on the control pilot (CP) line jumps to about 9 volts. (Zoomed in).
 3	This section shows the start of bidirectional communication between the vehicle and the EVSE, lasting for about 6 seconds. This communication uses a pulse-width modulation with a frequency of around 1,000 Hz and an amplitude modulation of between +9 volts and -12 volts. (Zoomed in).

 It is important to remember that when an EV is connected to the electric vehicle supply equipment (EVSE) to charge, the high-voltage vehicle system will be live and awake. Therefore, all high-voltage precautions will need to be observed, including the use of correctly rated and calibrated electrical test equipment and high-voltage Personal Protective Equipment (PPE).

In order to conduct any in-vehicle testing of mains charging systems using an oscilloscope, a differential probe with a minimum of a Category 3 (CAT III) rating will be required. A standard attenuator is not suitable and does not provide the correct protection for both the operator and the scope or vehicle.

Mastering Signal Analysis

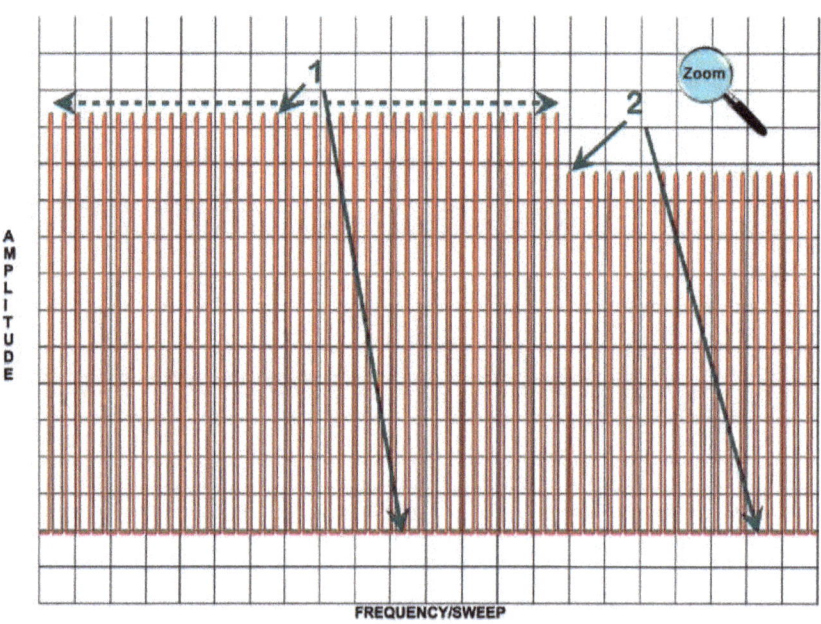

Figure 5.26 CP charging start (zoomed)

Table 5.11 CP connected charging start (zoomed)

Waveform component	Description
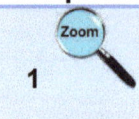 1	This section shows the start of bidirectional communication between the vehicle and the EVSE, lasting for about 6 seconds. This communication uses a pulse-width modulation with a frequency of around 1,000 Hz and an amplitude modulation of between +9 volts and -12 volts. (Zoomed in).
 2	This point on the waveform shows where charging has started, and the voltage drops to a modulated amplitude between +6 volts and -12 volts. Bidirectional communication continues and the duty cycle of the pulse width varies with the current delivered. (Zoomed in). (*See Table 5.13*).

When diagnosing electric vehicle (EV) charging issues, a multi-layered approach enhances accuracy. By synchronising oscilloscope readings with live data from a scan tool and dashboard indications, you gain a complete picture of the charging system's health.

- Capture the Control Pilot: Use an oscilloscope to inspect signal integrity during the handshake process between the vehicle and charging station. Verify the expected PWM duty cycle for current limits and ensure clean voltage transitions.
- Retrieve Live Data from the Scan Tool: Access charging system PIDs (Parameter IDs) such as state of charge (SOC), charging voltage, current draw, and control pilot interpretation. Compare scan tool readings with oscilloscope measurements—any mismatch may indicate faulty wiring or miscommunication between modules.
- Cross-Check with Dashboard Indications: Confirm that the vehicle's charge indicator aligns with oscilloscope and scan tool data. Investigate warning lights or fault messages that might suggest overvoltage, undervoltage, or pilot signal errors.

Mastering Signal Analysis

Table 5.12 indicates the charging states from the EVSE depending on approximate voltage amplitude.

Table 5.12 EVSE charging state

Charging cable connected	Charging status	Approximate voltage	Charging Possible	Description
No	Standby	0 volts	No	EVSE not connected to the vehicle.
Yes	Vehicle connected	9 volts	No	EVSE connected to the vehicle but not yet charging.
Yes	Charging	6 volts	Yes	EVSE connected to the vehicle, charging in progress.
Yes	Ventilation	3 volts	Yes	Ventilation in progress, charging fans running.
Yes	EVSE Shutdown	0 volts	No	EVSE fault or proximity pilot (PP) short to Earth.
Yes	Error	-12 volts	No	EVSE unavailable.

Duty cycle of the PWM signal and approximate current delivery are shown in **Table 5.13**.

Table 5.13 Relationship between duty cycle and charge current

Duty cycle percentage (%)	Approximate charge current
10%	6 Amps
20%	12 Amps
30%	18 Amps
40%	24 Amps
50%	30 Amps
66%	40 Amps
80%	48 Amps
90%	65 Amps
94%	75 Amps
96%	80 Amps

Motor controllers

Motor controllers regulate the power supplied to the electric motor. An electronic circuit is created so that four switches (in this case **Insulated Gate Bipolar Transistors** IGBTs) can change the route of electric current through a load, alternating the flow from one direction to another as shown in **Figures 5.27**.

For simplicity, the IGBTs are shown as open or closed switches in the following illustrations.

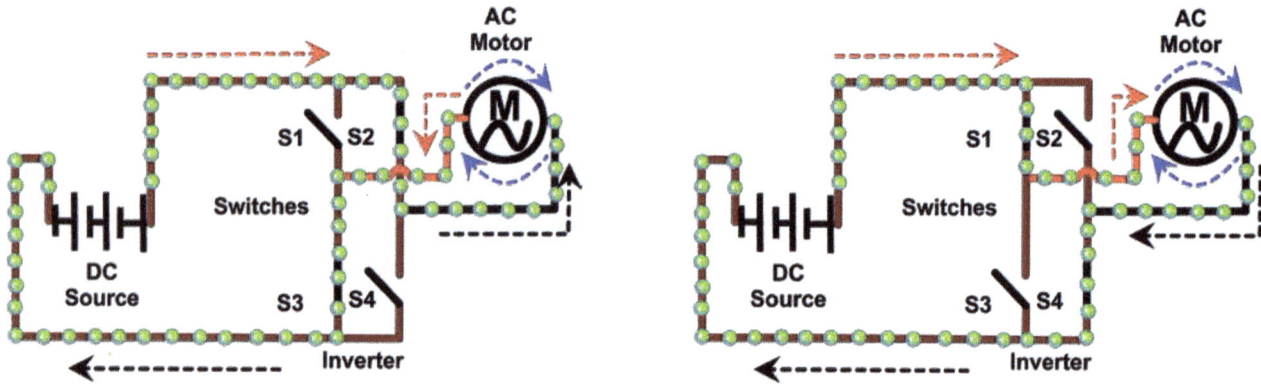

Figure 5.27 Changing current direction using switches

Although this is effectively alternating the current flow through the circuit, simplistic switching such as this will create a very crude square wave and not the smooth sinewave that is needed to efficiently operate the vehicle motor drive system.

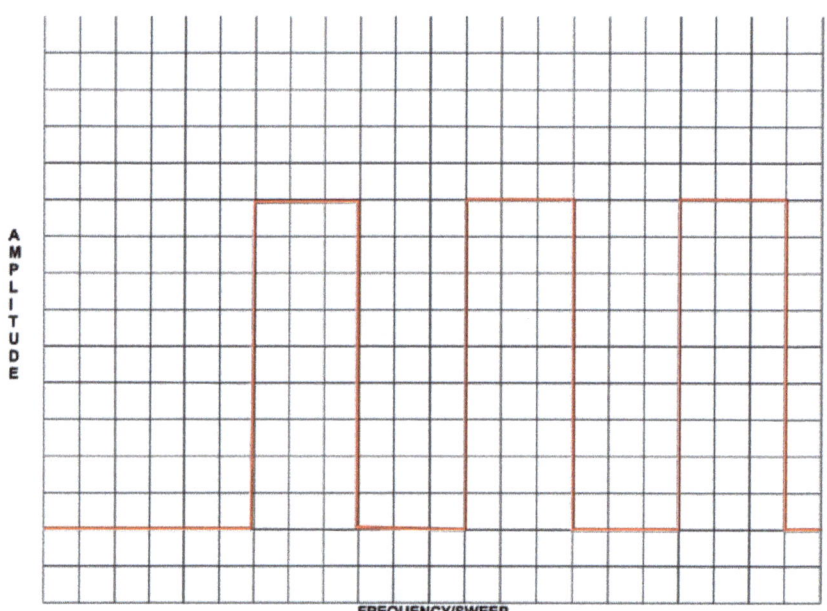

Figure 5.28 A square wave

Mastering Signal Analysis

 **Insulated Gate Bipolar Transistor IGBT** - a semiconductor device used in vehicle power electronics, combining high efficiency with fast switching capabilities. It acts as an electronic switch to control and convert power in systems such as electric vehicle inverters and motor drives.

If a controller is used to modulate the switching pulse time by rapidly opening and closing the switches multiple times per cycle, an average rising and falling voltage can be obtained. This closely resembles a smooth sinewave.

Figure 5.29 Modulated switching

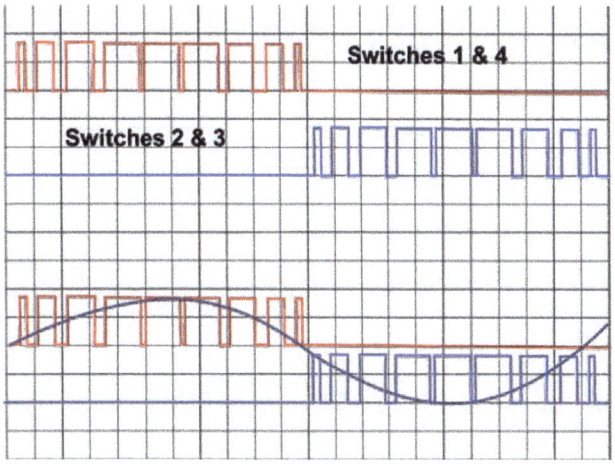

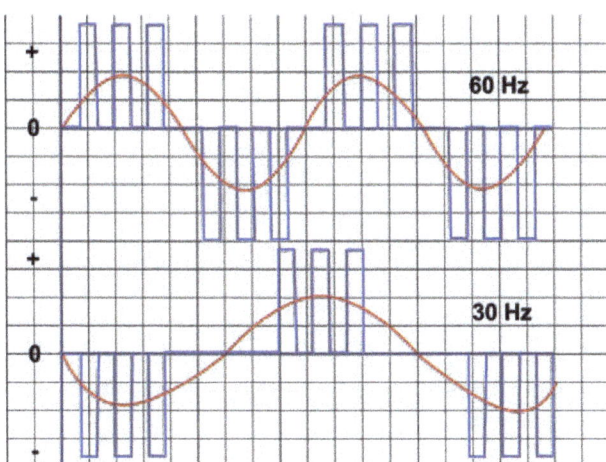

The inverter circuit is also able to vary the output voltage by controlling how long the switches are closed and current is flowing using pulse width modulation (PWM).

Figure 5.30 Voltage control with PWM

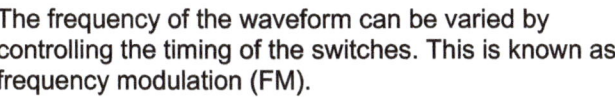

The frequency of the waveform can be varied by controlling the timing of the switches. This is known as frequency modulation (FM).

Figure 5.31 Frequency modulation

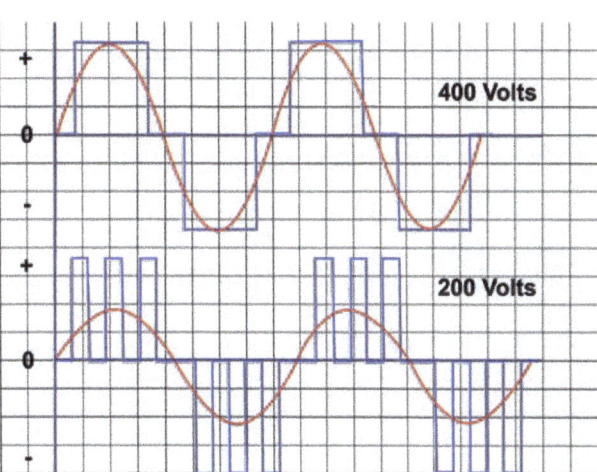

Mastering Signal Analysis

To test motor controllers.

Setup and use

Step 1: Observing all high-voltage health and safety precautions, including the use of HV PPE, shutdown and isolate the high-voltage system.

Step 2: Safely raise and support the vehicle to allow the motor and wheels to be driven when tested.

Step 3: Gain access to the motor generator 3-phase power cables and install specialist breakout box.

Step 4: Set up the oscilloscope to measure the motor voltage with a correctly rated differential probe connected to the breakout box.

Step 5: Power up the high voltage system and drive the electric motor(s).

Step 6: Monitor the motor controller outputs during various operating conditions.

Step 7: Look for consistent and stable waveforms that indicate proper control. The results will vary between manufacturers, so the waveform should be referenced to a 'known good'.

Figure 5.32 Testing motor controllers

Mastering Signal Analysis

Injector performance testing

Fuel injectors are critical components used with internal combustion engines, ensuring precise delivery of fuel into the combustion chamber for both petrol and diesel engines. Proper testing of these injectors is essential for maintaining engine performance and efficiency. This section describes the use of automotive oscilloscopes to analyse the injection duration and operation for both petrol and diesel engines, using waveforms to provide insights into their performance.

Petrol fuel injectors

Petrol fuel injectors operate under high pressure and short duration bursts, delivering a fine mist of fuel for optimal combustion. Using an oscilloscope, you can capture waveforms that reveal the operation of the injectors.

Waveform analysis

When analysing petrol fuel injectors, the oscilloscope captures a series of voltage spikes that indicate the injector's opening and closing events.

The waveform typically includes:

- Pull Down Voltage: Most petrol injectors switch to ground, pulling the signal down close to zero; this signifies the injector's solenoid activation, causing the injector to open.
- Injection Duration: The period during which the voltage remains low represents the injector being open and fuel being sprayed. This duration is crucial for determining the precise amount of fuel delivered.
- Closing Event: A secondary spike (back EMF) represents the injector's solenoid de-energising, allowing the injector to close and cease fuel delivery.

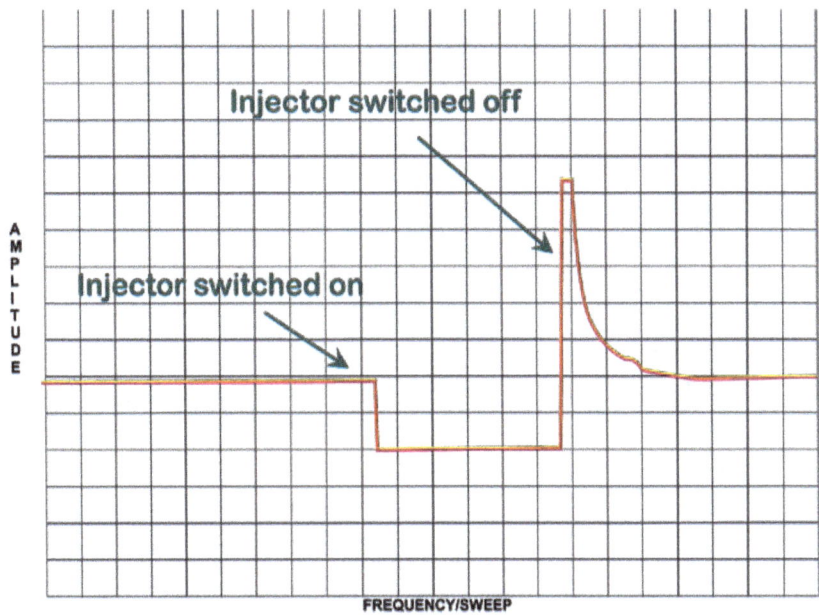

Figure 5.33 Petrol injector voltage waveform

Interpreting the waveforms

By examining these waveforms, you can determine if the injectors are operating within the specified parameters. Deviations in peak voltage, injection duration, or the shape of the waveform can indicate issues such as electrical faults, or mechanical wear.

When diagnosing petrol fuel injectors using an automotive oscilloscope, it's important to distinguish between the injector's electrical activity and its mechanical function. The waveform generated reflects the injector's switching behaviour—how voltage or current are applied to open and close the solenoid—but does not directly represent the fuel flow or spray pattern.

Diesel fuel injectors

Diesel injectors operate at much higher pressures compared to petrol injectors and use solenoid, piezoelectric or electromechanical Pumpe-Düse actuators to control fuel delivery. Analysing diesel injectors with an oscilloscope provides insights into their complex operation. Due to their design and operation, most diesel fuel injectors are best tested by connecting the oscilloscope to measure current, using an amp clamp.

When operated, solenoid diesel injectors may produce slightly different waveforms depending on their design. The following section shows some waveform patterns that might be displayed.

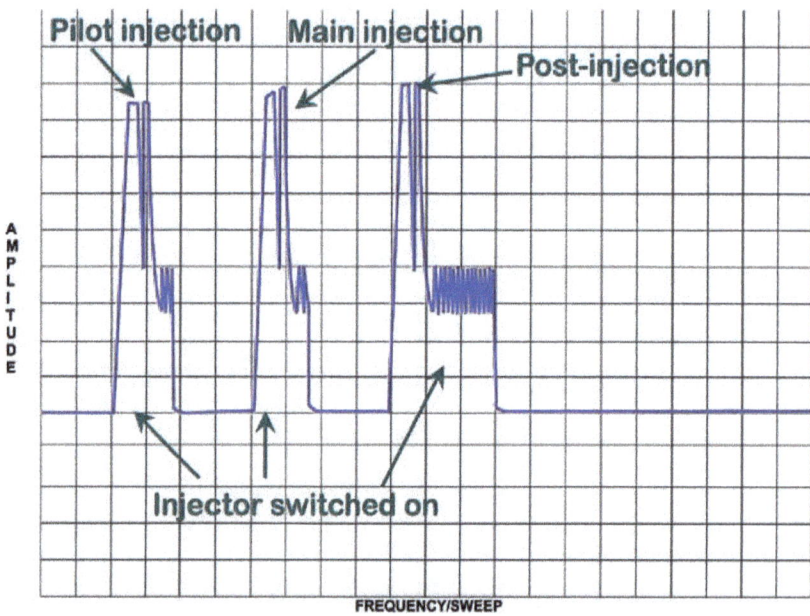

Figure 5.34 Common rail diesel solenoid injector current waveform

Mastering Signal Analysis

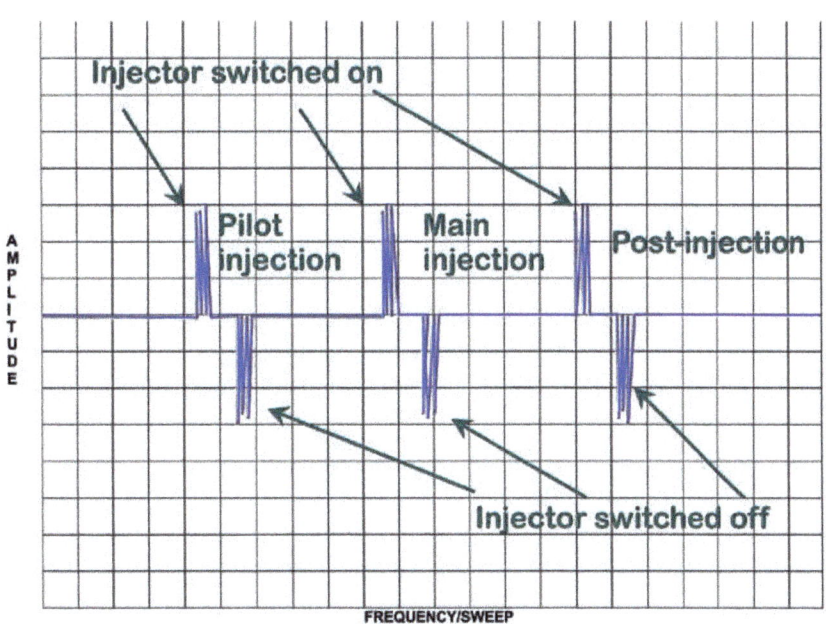

Figure 5.35 Common rail diesel piezoelectric injector current waveform

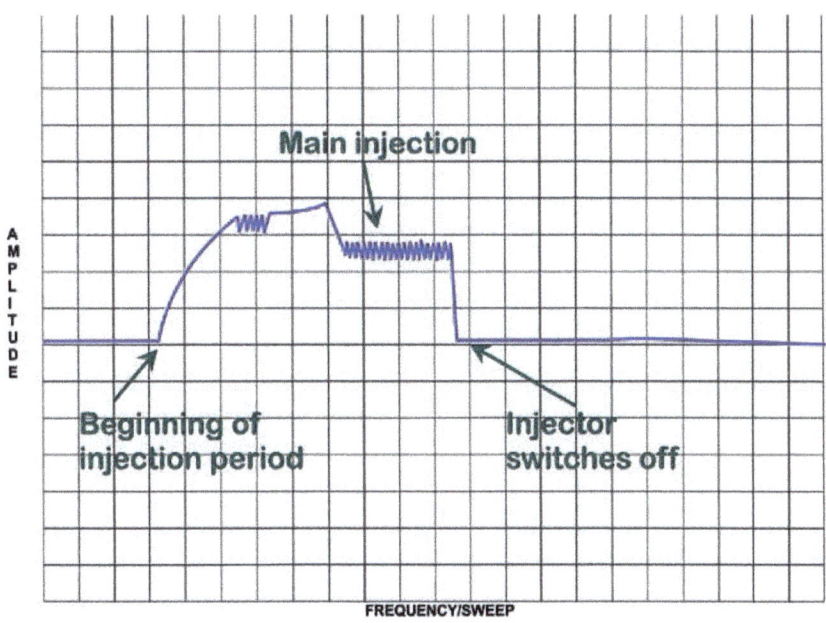

Figure 5.36 Pumpe-Düse injector current waveform

Mastering Signal Analysis

Waveform analysis

The waveform for diesel injectors is more intricate due to the multiple injection events (pilot, main, and post-injections).

The waveform can be broken down into:

- Pilot Injection: An initial current spike that introduces a tiny amount of fuel to begin combustion smoothly.
- Main Injection: The primary injection event, represented by a larger and longer spike in current and delivers the bulk of the fuel.
- Post-Injection: A final additional current spike following the main injection that assists in emissions control.

Interpreting the waveforms

The diesel injector waveform provides valuable data on the timing, duration, and sequence of the injection events. You can use this data to diagnose issues such as incorrect timing, or faulty actuators.

Practical steps for testing

Follow these steps to set up and test fuel injectors using an oscilloscope.

Setup and use

Step 1
- Setup: Connect the oscilloscope probes to the injector signal wire and ground for voltage testing. Use appropriate breakout leads and attenuators if necessary. Use a current clamp for diesel injector testing.

Step 2
- Configuration: Configure the oscilloscope to capture the waveform at a high resolution. Set appropriate voltage or current, and time scales to visualise the injection events clearly.

Step 3
- Monitoring: Observe the waveforms during various operating conditions like idling, acceleration, and deceleration to ensure consistent and stable operation.

Step 4
- Analysis: Compare the captured waveforms against known good patterns to identify any anomalies. Pay close attention to the injection duration, peak voltage, and the shape of the waveform.

Ignition system analysis

The ignition system is crucial for the proper operation of an internal combustion engine. By using an oscilloscope to capture and analyse ignition coil and spark plug signals, you can diagnose and resolve various engine performance issues.

Setup

To accurately capture ignition signals, connect the oscilloscope probes to the primary and secondary ignition circuits. For capturing the primary circuit, connect the probe to the ignition coil's primary winding, typically connected to the ignition control module or ECU. For the secondary circuit, use a capacitive or inductive pickup around the spark plug wire or a wand probe.

Mastering Signal Analysis

Figure 5.37 Secondary ignition wand probe

Configuration

Configure the oscilloscope for high-resolution waveform capture. Set the voltage scale appropriately for the primary and secondary circuits and adjust the time base to capture the entire ignition event, including the dwell time, spark duration, and any secondary oscillation (sometimes called **ringing**).

Monitoring

Observe the ignition waveforms under different engine operating conditions such as idling, acceleration, and under load. The primary ignition waveform should show the **dwell period** and the voltage spike as the coil discharges. The secondary waveform should depict the spark voltage, spark duration, and any post-spark oscillations.

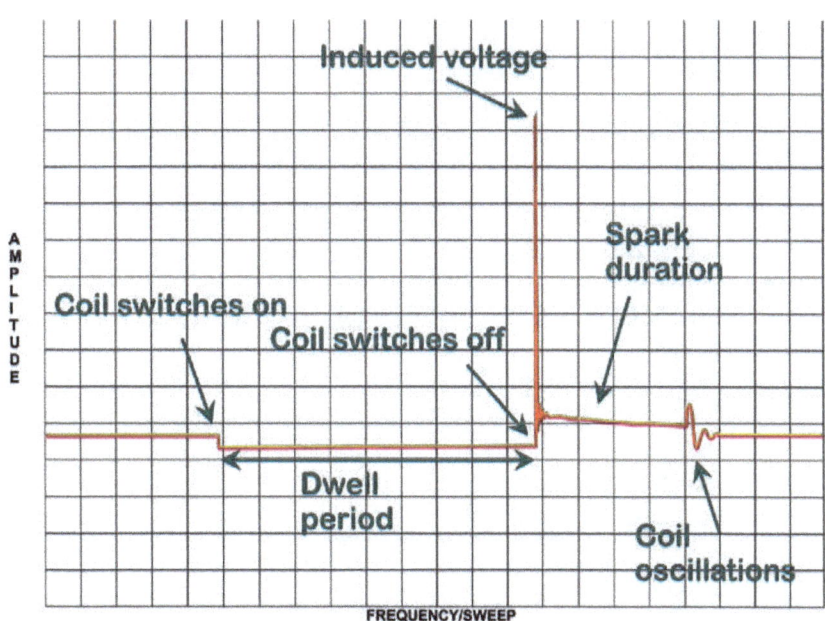

Figure 5.38 Ignition primary coil waveform

Mastering Signal Analysis

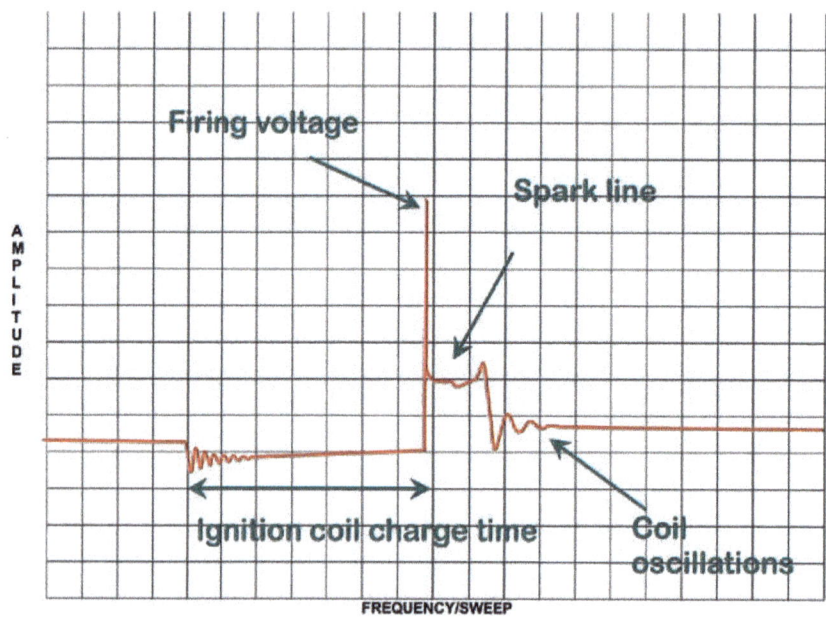

Figure 5.39 Ignition coil secondary waveform

Analysis

Compare the captured waveforms with known good patterns to identify issues. Pay attention to the following parameters:

- Dwell Time: The period during which the ignition coil is charging. Short or erratic dwell times can indicate problems with the ignition control module or coil driver circuits.
- Primary Voltage Spike: The peak voltage during coil discharge. A low or absent spike can suggest a weak coil or grounding issues.
- Spark Voltage: The peak voltage required to jump the spark plug gap. Excessively high or low spark voltage may indicate worn spark plugs, incorrect gap, ignition coil, compression or air/fuel ratio issues.
- Spark Duration: The time the spark is maintained across the plug gap. Short spark duration can result from a weak ignition coil or poor spark plug condition.
- Secondary Ringing: Oscillations following the spark event, which can provide clues about the condition of the ignition coil and spark plug.

Ringing - oscillations or transient disturbances in an electrical signal, often caused by inductive components, impedance mismatches, or sudden changes in voltage.

Dwell period - the time during which current flows through the ignition coil's primary winding to build up magnetic energy before it is released as a spark in the combustion chamber.

Alternator and charging system diagnostics

The alternator is responsible for maintaining the battery's charge and supplying electrical power to the vehicle's systems while the engine is running. Diagnosing alternator and charging system issues is essential for ensuring the reliability and efficiency of a vehicle. One of the most effective methods for diagnosing these issues is by evaluating **ripple** voltage and current waveforms using an automotive oscilloscope.

Mastering Signal Analysis

Understanding ripple voltage

Ripple voltage is the residual alternating current (AC) component found within the direct current (DC) output of the alternator. The presence of ripple voltage is a normal characteristic of the alternator's operation due to the rectification process, where AC is converted to DC. However, excessive ripple voltage can indicate problems within the alternator or the charging system.

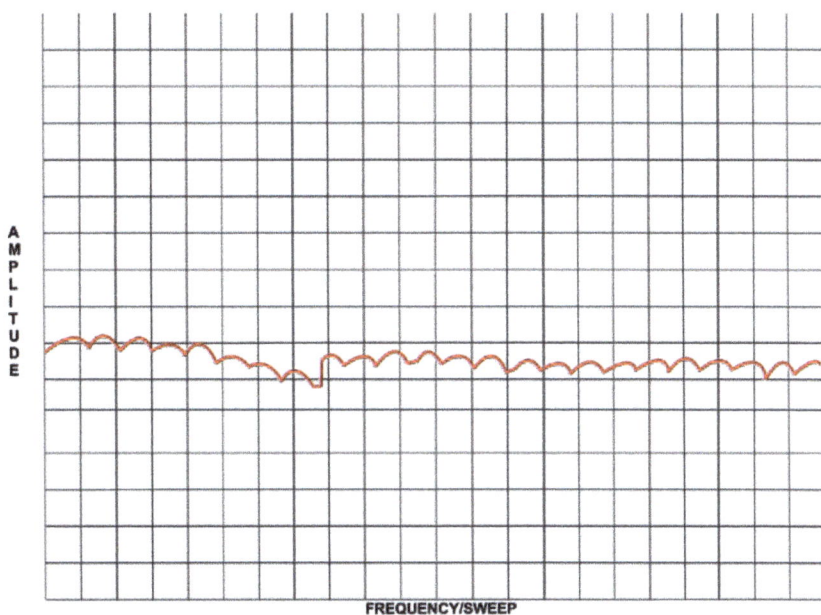

Figure 5.40 Alternator rectified voltage

Causes of excessive ripple voltage

- Faulty Diodes: The alternator contains **diodes** that **rectify** the AC output to DC. If one or more diodes fail, excessive ripple voltage can result.
- Worn Brushes: Worn **brushes** can lead to poor electrical contact within the alternator, resulting in increased ripple voltage.
- Damaged Windings: Damaged or shorted windings within the alternator can disrupt the electrical output, causing higher ripple voltage.
- Poor Connections: Loose or corroded connections in the charging system can contribute to abnormal ripple voltage levels.

Ripple - the residual alternating current (AC) voltage present in the output of an alternator, caused by incomplete rectification of AC to direct current (DC).

Diode - a semiconductor device that allows current to flow in one direction while blocking it in the opposite direction.

Mastering Signal Analysis

Rectify - to convert alternating current (AC) into direct current (DC) using electronic components like diodes or rectifier circuits, ensuring unidirectional flow of electrical energy.

Brushes - conductive components made of carbon or graphite that maintain contact with the rotating commutator or slip-rings in devices like alternators or starter motors, enabling electrical current transfer.

Testing ripple voltage with an oscilloscope

Setup and use

- **Step 1**: Connect the oscilloscope probes to the battery terminals, ensuring correct polarity.
- **Step 2**: Set the oscilloscope to AC voltage mode to filter out the DC component, focusing on the ripple voltage.
- **Step 3**: Start the engine and observe the waveform on the oscilloscope.
- **Step 4**: Compare the captured waveform with known good patterns to identify any abnormalities. A healthy alternator will show a consistent waveform with minimal ripple voltage. Excessive peaks or erratic patterns indicate potential issues that require further investigation.

Additional considerations

When diagnosing alternator and charging system issues, it's essential to consider the following:

- Battery Condition: Ensure the battery is in good condition and fully charged before testing the alternator and charging system.
- Engine Load: Test the alternator under various engine loads to assess its performance accurately.
- Temperature: Consider the impact of temperature on the alternator's performance, as extreme temperatures can affect its efficiency.
- Vehicle Electrical Demands: Evaluate the charging system while operating various electrical components (e.g., headlights, air conditioning) to ensure the alternator can meet the vehicle's demands.

Pressure transducers

Pressure transducers are invaluable tools when used in automotive diagnostics, particularly for in-cylinder compression testing, inlet and exhaust pressure pulse analysis. These devices convert pressure into an electrical signal, which can then be visualised on an oscilloscope, providing valuable insights into engine health and performance.

Mastering Signal Analysis

In-cylinder compression testing

In-cylinder compression testing using pressure transducers allows you to precisely measure the pressures within an engine cylinder throughout the entire engine cycle. By connecting a pressure transducer and cranking or running the engine, the sensor captures the pressure variations which can then be analysed. The resulting waveform on the oscilloscope reveals the peak compression pressure, induction troughs and exhaust backpressure which can be compared to manufacturer's specifications to determine the condition of the engine's internal components such as pistons, rings, and valves.

With a dedicated adapter attached to the pressure transducer, and the oscilloscope connected to a petrol engine at the spark plug fitting, the pulsations shown on the display can then be used to perform a compression diagnosis of a single cylinder during engine operation.

To test the cranking compression, the engine must be isolated so that it doesn't start.

A mechanical/pneumatic compression gauge includes a one-way valve in the connection adapter. When the engine is cranked, the displayed value is the result of multiple compression cycles combined.
In-cylinder pressure transducers offer a more precise, real-time measurement of compression pressures. The waveform shows each individual compression event, enabling better analysis of the engine's mechanical condition.

A pressure transducer adapter should only be used/connected to recommended systems that operate within the manufacturing tolerance of the equipment.
Exceeding the measurable/recommended pressures can lead to personal injury, vehicle and equipment damage.

It is also very important to remember that connecting a pressure transducer to a cylinder for in-cylinder compression testing will cause that cylinder not to operate in the normal manner. The misfire created by removing the spark plug and inserting the adapter could lead to catalytic converter damage. The <u>ignition and fuel injection MUST be isolated</u> cranking compression testing to reduce the possibility of damage or injury. For running tests, a spark plug simulation may be necessary to reduce the possibility of ignition coil damage.

It is recommended that any in-cylinder compression testing is conducted for the shortest possible amount of time required to gather diagnostic information.

Running the engine with a spark plug removed will normally cause a diagnostic trouble code to be generated, and the malfunction indicator lamp (MIL) may illuminate. It is important to ensure that all diagnostic trouble codes have been cleared after any work has been conducted.

Mastering Signal Analysis

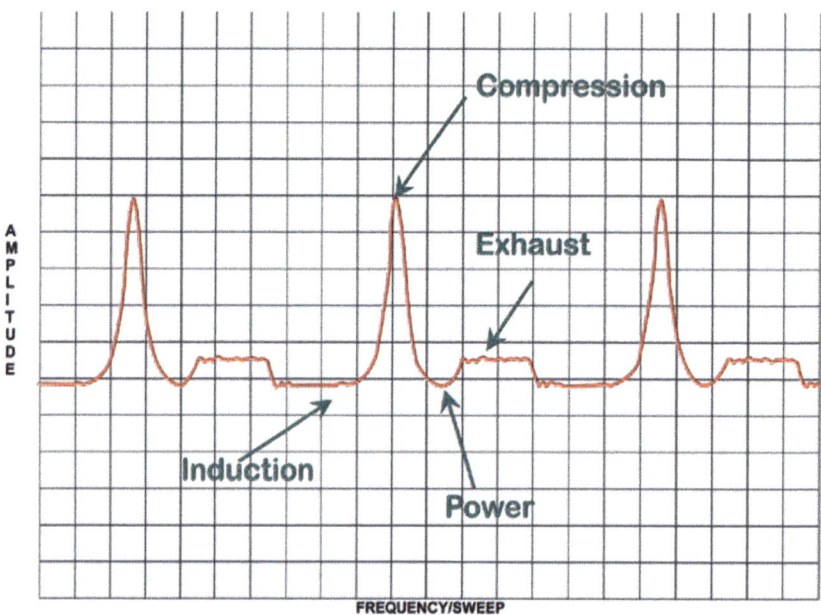

Figure 5.41 In-cylinder compression pressures with the engine running

Inlet and exhaust pressure pulse analysis

Inlet and exhaust pressure pulse analysis is another application of pressure transducers. By installing pressure transducers in the intake and exhaust systems, you can monitor the pressure pulses generated by the engine during operation. These pulses provide insights into the airflow dynamics and the performance of the intake and exhaust valves. On the oscilloscope, a healthy system displays regular and consistent pressure pulses, while abnormal patterns may indicate issues such as valve timing problems, restrictions in the intake or exhaust systems, or exhaust backpressure issues.

Analysing intake and exhaust pressure pulses using an oscilloscope is much like a doctor evaluating a patient's breathing patterns. Just as a physician listens to inhalation and exhalation for signs of respiratory health, you can interpret pressure waves moving through the engine's air system to diagnose combustion efficiency, valve timing, and airflow restrictions.

The Engine as a Respiratory System
- Intake Stroke → Inhalation
- Compression Stroke → Breath Holding
- Power Stroke → Controlled Release
- Exhaust Stroke → Exhalation

Just as a doctor uses a spirometer to visualise breath flow in a lung function test, an oscilloscope helps you identify irregular pressure waves caused by faulty valve sealing, blocked air passages, or timing errors. A smooth intake and exhaust waveform suggests optimal airflow, while distorted, delayed, or missing pulses could indicate mechanical problems, much like identifying a respiratory disorder.

Mastering Signal Analysis

Petrol intake manifold pressure

With the pressure transducer connected to the oscilloscope, adapters can be used to attach it to the intake manifold of a petrol engine via a suitable vacuum pipe connection; this should be after the throttle butterfly. With the engine being cranked, the intake pulses can then be analysed.

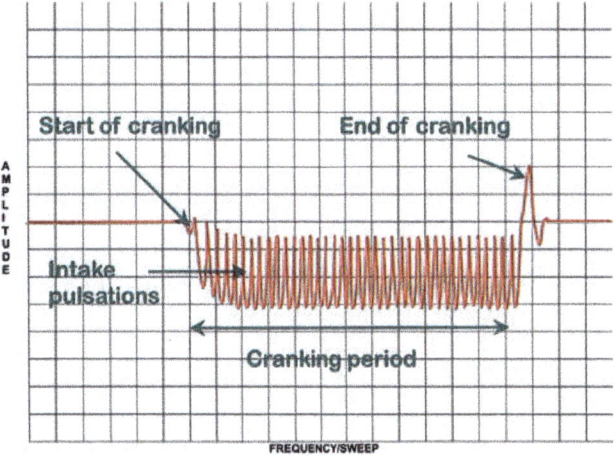

Figure 5.42 Intake manifold pressure while cranking

With the pressure transducer still connected to the intake manifold and the engine running, the throttle can be rapidly opened and closed, with the response checked using the waveform. If a relatively slow time scale/sweep is selected, this will give you the opportunity to evaluate the intake manifold pressure under a number of different operating conditions:

- Idle
- Acceleration
- Wide open throttle (WOT)
- Engine off

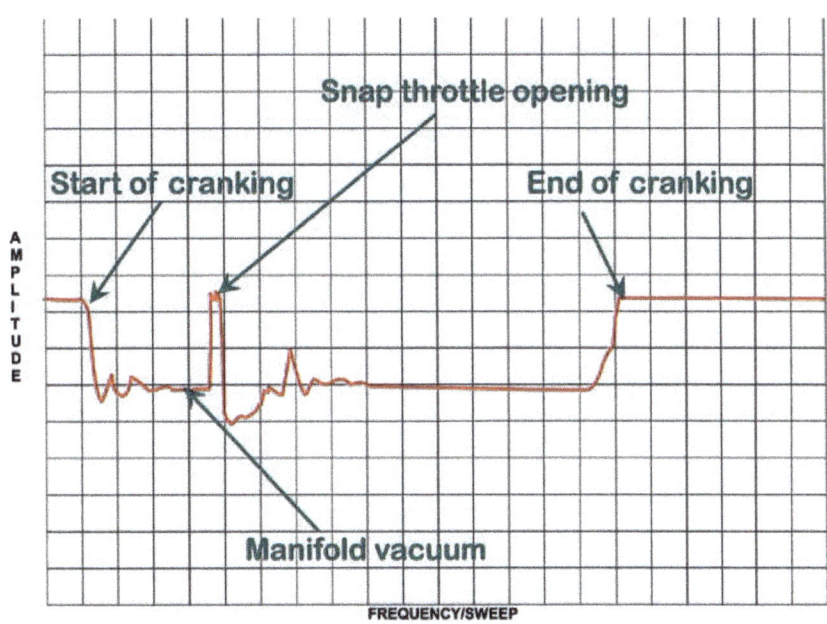

Figure 5.43 Intake manifold pressure with snap open throttle

Exhaust pulses

With an adapter attached to the pressure transducer, the oscilloscope may also be connected to the vehicle at the exhaust tail pipe and the pulsations shown on the display can then be used to perform a mechanical diagnosis of engine operation.

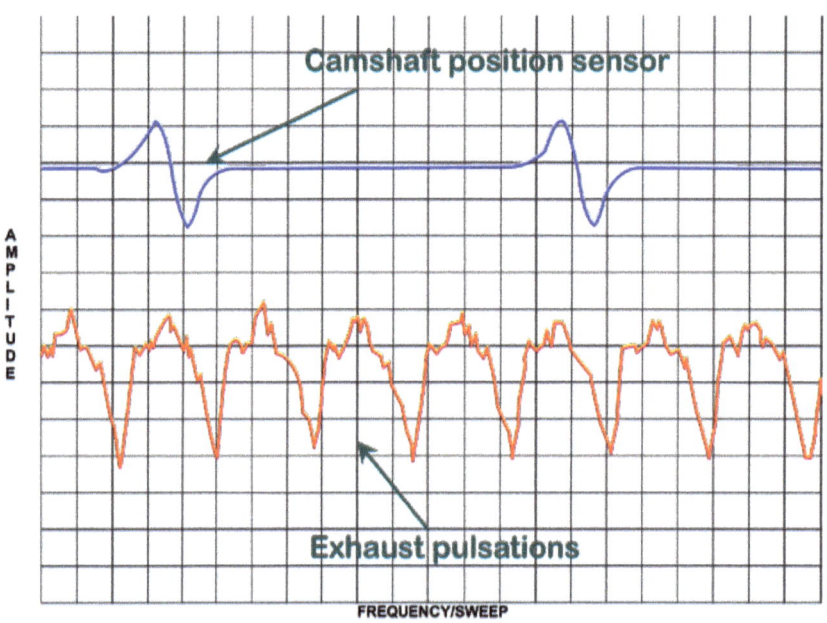

Figure 5.44 Exhaust tailpipe pressure with engine cranking

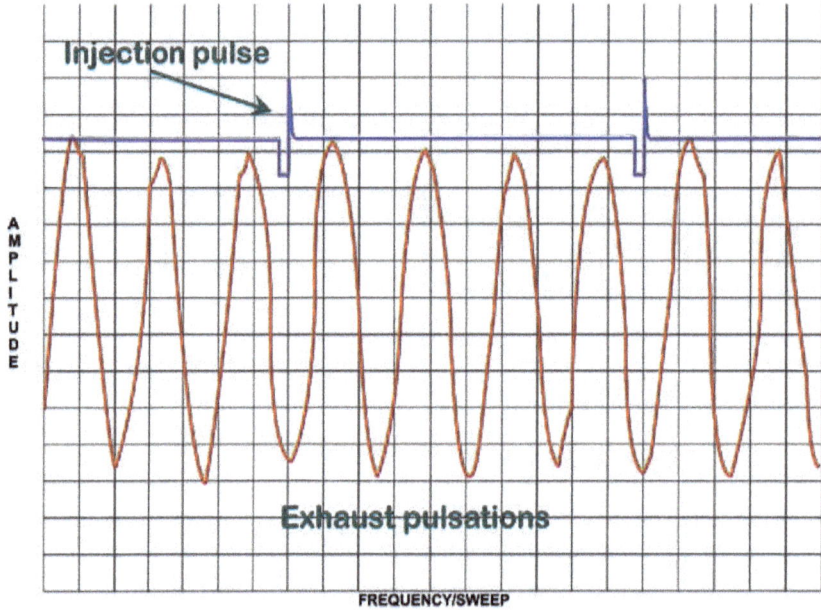

Figure 5.45 Exhaust tailpipe pressure with engine running

Mastering Signal Analysis

Connecting a secondary channel to a signal source, such as the camshaft sensor or fuel injector, provides a timing reference that can assist in identifying cylinders exhibiting anomalies during intake or exhaust pressure pulse analysis.

Noise vibration and harshness NVH

Noise, vibration, and harshness (NVH) are critical factors in automotive diagnostics and maintenance, affecting the comfort, performance, and longevity of vehicles. NVH issues can be indicative of underlying mechanical problems and can significantly impact the driving experience. Automotive oscilloscopes serve as invaluable tools for technicians in diagnosing and resolving NVH-related issues, providing a detailed analysis of various vehicle systems.

Using oscilloscopes for NVH testing

Oscilloscopes, equipped with specialised adapters, allow you to measure and analyse the vibrations and noise produced by different parts of a vehicle. These adapters can be attached to various components, such as the engine, transmission, chassis, suspension and other key areas, to capture vibration pulsations and changes. By interpreting the visual data displayed on the oscilloscope, you can diagnose issues with precision.

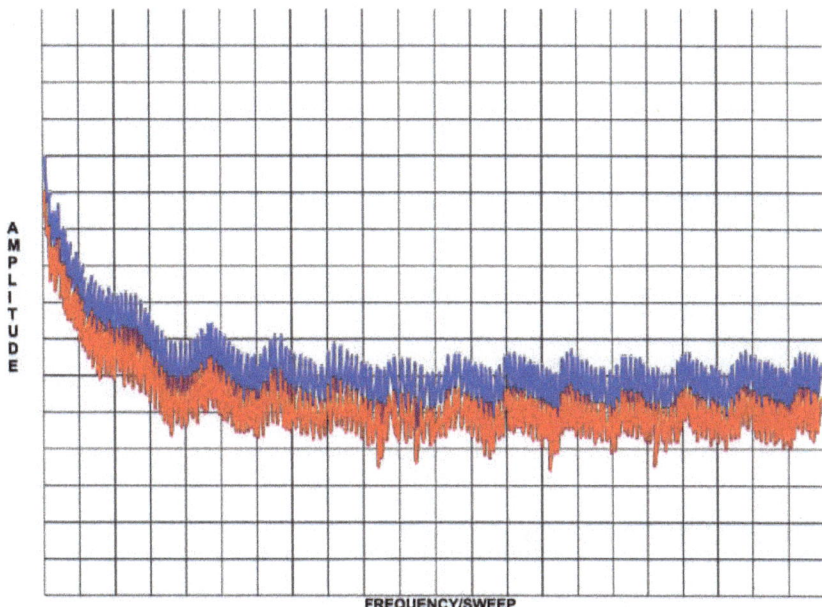

Figure 5.46 NVH waveform

Conclusion

After mastering the operation and functions of an oscilloscope, precise signal analysis is essential for accurate diagnosis. It is important to compare test results to known good examples where possible. Generally, there are a limited number of correct examples but countless incorrect possibilities.

Chapter 6 Advanced Techniques

As technology continues to advance, the role of the automotive technician has become increasingly sophisticated and demanding. Oscilloscopes, essential tools in the modern mechanic's toolkit, provide a detailed view into the electronic and electrical systems of vehicles, facilitating precise diagnostics and repair. With a solid understanding of the foundational aspects of setup and operation, it is now possible to delve into more advanced techniques that will enhance diagnostic strategies and routines to a higher level of expertise.

Advanced techniques using automotive oscilloscopes requires a deeper comprehension of signal processing, custom profile creation, and dynamic testing. These skills enable you to identify issues that may not be immediately apparent through basic diagnostics.

Contents

Custom Signal Capture Profiles	171
Setting Up Custom Probes	172
Dynamic Testing Under Load	175
Measurements	181
Masks	184
Actions	186
Reference Waveforms	188
Maths Channels	191
Error Analysis in Complex Networks	199
Case Studies	205
Educational Insights	208
Common Missteps and Troubleshooting Tips	208
Developing Diagnostic Strategy	209
Future Technologies	210
AI Integration in Oscilloscope Diagnostics	210
Next-Generation Communication Protocols	210

Advanced Techniques

The automotive industry is a high-risk environment, especially when dealing with electrical systems. The hazards of electricity are well-known but can be easily ignored due to its invisible nature. This can lead to complacency if the fundamentals of electricity are not well understood. Even with this understanding, caution is necessary. Assume that any safety systems designed for protection have failed and take precautions to minimise the risk of injury or death. Always evaluate the risks associated with any activity and implement measures to eliminate or reduce the hazards involved in any task, diagnosis, or repair. Additional risks associated with working on, or around electrical systems may include:

- Electrocution
- Strong magnetic fields
- Falling from heights
- Short circuits
- Electrical discharge/arcing
- Fire and explosion
- Chemicals

Custom Signal Capture Profiles

Automotive fault-finding and repairs require precision and accuracy, particularly when diagnosing electrical systems. Custom signal capture profiles are beneficial, as they enable repeated diagnostics tailored to specific vehicles or systems.

Understanding custom signal capture profiles

Custom signal capture profiles are predefined settings within an oscilloscope that allow you to consistently capture and analyse the same types of signals across different diagnostic sessions. These profiles streamline the process, ensuring that each diagnosis is conducted under the same parameters, which can be crucial for accurate comparison and analysis over time.

Benefits of custom signal capture profiles

- Consistency: By using custom profiles, you can ensure that each diagnostic session starts with the same parameters, reducing variability and increasing reliability.
- Efficiency: Predefined profiles save time by eliminating the need to manually set up the oscilloscope for each session.
- Accuracy: Consistent settings allow for precise comparison between diagnostics, highlighting changes or anomalies in the system being tested.

Using custom profile settings on an automotive oscilloscope is like organising a toolbox for rapid repairs.
Imagine walking into a workshop where every tool is scattered. Before making a single fix, you waste time searching for the right spanner, screwdriver, or socket. But with a properly arranged toolbox, each tool is already categorised, ready for immediate use—saving time and ensuring precision.
Likewise, when you custom preset oscilloscope profiles, your voltage ranges, time bases, trigger settings, and signal filters are all optimised for your diagnostic needs. Whether working on fuel injectors, CAN BUS signals, or EV charging systems, your scope is immediately ready—no need for repetitive adjustments. It's the difference between diving straight into problem-solving versus fumbling through setup before diagnosing a fault.

Advanced Techniques

Designing custom signal capture profiles

Creating custom signal capture profiles involves several steps, each requiring careful consideration to ensure that the profile captures the necessary data accurately and efficiently.

Setup and use

Step 1
- Identify the Target System: Begin by determining the specific vehicle or system you wish to diagnose. Understanding the subtleties of the target system is essential in designing an effective profile. This includes knowing the type of signals you will encounter, and the key components involved.

Step 2
- Define the Parameters: Set the parameters for signal capture based on the characteristics of the target system. These parameters may include:
- Time base: The duration over which the signal will be captured.
- Voltage range: The range within which the signal will be measured.
- Trigger settings: Conditions under which the oscilloscope starts capturing data.

Step 3
- Configure the Oscilloscope: Input the defined parameters into the oscilloscope. This involves setting the time base, voltage range, and trigger settings as identified in the previous step. Ensure that the oscilloscope is calibrated and ready for accurate data capture.

Step 4
- Test and Refine the Profile: Conduct a test run to verify the effectiveness of the custom profile. Review the captured data for accuracy and consistency. Make adjustments as needed to refine the profile, ensuring that it meets the desired diagnostic standards.

Step 5
- Save and Document the Profile: Once satisfied with the custom profile, save it to the oscilloscope for future use. Document the profile settings and the diagnostic process for reference. This documentation is vital for consistent application and for sharing the profile with colleagues.

Implementing custom signal capture profiles

With the profile designed and saved, implementing it in day-to-day diagnostics becomes straightforward. When a vehicle or system requires diagnosis, simply load the custom profile and begin capturing data. The consistency of the profile ensures that each session gives you comparable results, facilitating accurate analysis and troubleshooting.

Setting up custom probes

Custom probes allow you to tailor your diagnostic tools to specific requirements, enhancing precision and accuracy in capturing and analysing waveform data. This section looks into the process of setting up custom probes and explains the various measurement units that can be used to interpret the captured data.

Advanced Techniques

Step-by-step guide to setting up custom probes

Setup

Step 1
- Choose the right type of probe based on your diagnostic needs. Probes vary in their functionality and sensitivity, and selecting the appropriate one ensures that the data captured is relevant and accurate. Common types of probes include:
- Voltage Probes: Used to measure electrical potential difference. - Current Probes: Designed to measure the flow of electric current. - Temperature Probes: Used for measuring thermal variations. - Pressure Probes: Ideal for capturing pressure changes within systems.

Step 2
- Configure the settings on the oscilloscope. This includes the following steps:
- Input Parameters: Define the parameters such as time base, voltage range, and trigger settings.
- Calibration: Ensure the probe is calibrated correctly for accurate measurements. Calibration involves setting zero points and adjusting sensitivity. This will normally require access to the probe manufactures information and specifications.
- Connection: Securely attach the probe to the oscilloscope and ensure it has a stable connection.

Step 3
- Conduct initial tests to verify the effectiveness of the custom probe setup. Analyse the captured data for accuracy and consistency. Refine the settings as needed to ensure that the diagnostic standards are met. This step is crucial as it helps in optimising the probe's performance for various automotive diagnostics.

Step 4
- Once the probe setup is configured and tested, save the profile on the oscilloscope for future use. Document the settings and the process followed to create the profile. Thorough documentation aids in maintaining consistency and alows you to share the profile with colleagues for collaborative diagnostics.

Calibrating custom probes against other test tools will help define their accuracy. For example:

- A multimeter can be used to calibrate, voltage, current, resistance.
- A thermometer can be used to calibrate temperature.
- A pressure gauge can be used to calibrate transducers.

Implementing custom probes in diagnostics

With the probe setup saved and documented, using it in day-to-day diagnostics becomes straightforward. When a vehicle or system requires analysis, simply load the custom profile and begin capturing data. Consistent application of the profile ensures reliable results, aiding in accurate analysis and troubleshooting.

Figure 6.1 A custom probe

Advanced Techniques

How to setup a custom linear probe using Max/Min values

Obtain Manufacturer Specifications:
- Look up the sensor's output voltage range and the corresponding engineering units.
- For example: A pressure sensor might output 0.5V to 4.5V linearly for 0 psi to 150 psi.

Use These as Your Scaling Points:
- Minimum input voltage: 0.5V
- Maximum input voltage: 4.5V
- Corresponding output units: 0 psi (at 0.5V) to 150 psi (at 4.5V)

Input This into Your Scope's Probe Configuration:
- Choose "Custom Probe" or "User Defined Probe" in the oscilloscope's menu.
- Set probe type to linear.

Enter:
- Minimum voltage = 0.5V, value = 0
- Maximum voltage = 4.5V, value = 150

The software will then auto-calculate the linear slope and offset:

$$\text{Scale (Gain)} = \frac{150 - 0}{4.5 - 0.5} = 37.5 \text{ psi/V}$$

$$\text{Offset} = -18.75 \text{ psi}$$

(Because at 0V, it's below scale)

Label and Save the Probe:
- Give the probe a clear name (e.g., 'Pressure transducer 0-150 psi').
- Save it so you can recall it easily later.

Always verify the probe output by comparing it to known good data or a reference gauge to ensure calibration is accurate—especially if tolerances or sensor aging might affect output.

Understanding measurement units

A key aspect of using custom probes effectively is understanding the measurement units available and their relevance in automotive diagnostics.

Table 6.1 provides an explanation of various measurement units that can be utilised:

Advanced Techniques

Table 6.1 Measurement units

Unit	Measurement type	Relevance to diagnostics
Volts (V)	Measurement of electrical potential difference.	Essential for diagnosing electrical faults.
Amps (A)	Measurement of electric current flow.	Important for analysing current consumption and identifying where excess current occurs.
Watts (W)	Measurement of power.	Useful for assessing the efficiency of electrical components.
Seconds (s)	Time measurement.	Crucial for timing events and understanding durations.
Frequency (Hz)	Measurement of periodic occurrences per second.	Vital for understanding signal oscillations, including pulse width modulation and duty cycle.
Sound (dB)	Measurement of sound intensity.	Helpful in analysing levels of sound in relation to noise vibration and harshness (NVH).
Degrees (°)	Measurement of angles.	Used in analysing rotational displacement.
Rads (rad)	Measurement of angular displacement in radians.	Important for detailed angular measurements.
Ohms (Ω)	Measurement of electrical resistance.	Key for diagnosing resistance-related issues.
Baud	Measurement of data transmission rate.	Relevant for communication diagnostics.
s/div	Seconds per division.	Determines the scaling of the time base in oscilloscopes.
Percentage (%)	Measurement of proportions, ratios and provides comparisons.	Relative measurement often used in efficiency calculations and duty cycle.
Bar	A unit of barometric pressure.	Common in automotive pressure diagnostics.
psi	Pounds per square inch, a unit of pressure.	Widely used in air and fluid pressure measurements.
lpm	Liters per minute, a unit of flow rate.	Important for assessing fluid dynamics.
RPM	Revolutions per minute.	Critical for evaluating engine and component rotations.

Dynamic Testing Under Load

Dynamic testing under load shows how components perform in real operating conditions, unlike static testing which evaluates parts in isolation. It helps diagnose issues that don't appear in controlled scenarios.

When conducting dynamic testing, it is essential to consider the associated risks. The vehicle or system under test might have a fault, increasing the likelihood of failure, damage, or personal injury.
In the case of hybrid or electric cars, safety systems designed to protect against high voltage may be compromised, raising the risk of electrocution.
When operating engines, it is important to account for:
- Exhaust fumes
- Rotating or moving components
- Hot surfaces

Advanced Techniques

Realistic performance assessment

Dynamic testing under load provides a realistic evaluation of a component's performance. When exposed to real driving conditions, factors such as engine load, temperature changes, and varying pressures affect the behaviour of automotive parts. Recording waveforms in this environment shows the operational characteristics, allowing for precise diagnosis and adjustments.

Figure 6.2 Diagnostic road test

To conduct a road test, it is necessary to ensure the vehicle is roadworthy and complies with legal requirements for driving on a public highway. The driver must possess the appropriate licence and insurance and not be under the influence of drugs or alcohol. Avoid any activities while driving that distract from the act of driving.

Identification of intermittent faults

Static tests can provide limited information, often resulting in incomplete diagnostics. Some automotive issues are intermittent and only occur under specific conditions, making them elusive during static testing.
Dynamic testing allows you to observe how a component behaves during acceleration, deceleration, and varied load scenarios. This comprehensive capture helps in identifying faults that might otherwise go unnoticed, ensuring a thorough diagnostic process.

After preliminary checks, it is advisable to start with scanning the vehicle for diagnostic trouble codes (DTC). Ensure this scan covers all vehicle systems, sometimes referred to as a global scan. This approach is important because faults in interconnected systems may be the underlying cause of the issue being investigated.
It is also necessary to closely examine freezeframe data, which shows the vehicle's operating conditions when the diagnostic code was set. Although this data might not indicate when the problem started, it can provide useful information for reconstructing operating conditions during performance diagnostics or road tests.

Advanced Techniques

Methodology for dynamic testing under load

 Whenever possible, it is advisable to have at least two people conduct performance-related diagnosis. This method allows one person to focus on operating or road testing the vehicle, while the other focuses on data collection and the diagnostic process.

 To conduct dynamic oscilloscope testing, particularly during a road test, it is essential to use a portable oscilloscope that either has its own power supply or can be powered directly from the vehicle. Test probes and leads must be securely connected and routed in such a manner that they remain firmly attached throughout the test and do not interfere with other vehicle systems. Additionally, all doors, bonnets, tailgates, and similar components should be securely latched to prevent them from coming loose or opening during the road test.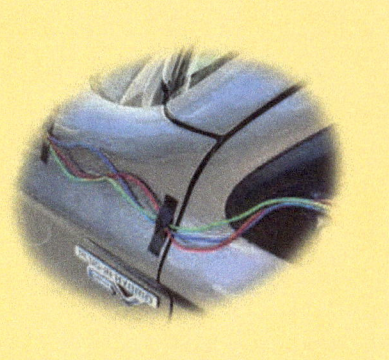

Preparation

Setup

Step 1
- Before initiating dynamic testing, ensure that the oscilloscope and all related equipment are correctly configured.

Step 2
- Verify connections to the components being tested.

Step 3
- Set the oscilloscope to appropriate settings, including time base (s/div), trigger points, and voltage range.

Step 4
- Use suitable probes and accessories to capture accurate waveforms.

Advanced Techniques

Vehicle setup

Setup

Step 1: Position the vehicle in a safe environment where it can be operated under load conditions. This could be on a dynamometer or a controlled road test route.

Step 2: Ensure all safety measures are in place to prevent accidents during testing, and suitable ventilation or exhaust extraction is used where necessary.

Step 3: Warm up the vehicle to its operating temperature for realistic assessment.

Oscilloscope setup

Setup

Step 1: Configure the oscilloscope to capture the necessary parameters.

Step 2: Set the time base (s/div) to align with the expected frequency of the signals being tested.

Step 3: Adjust the voltage range to accommodate the signals from the components.

Step 4: Use an appropriate maths channels to apply useful equations for deeper insight.

Notes: Many Digital Storage Oscilloscopes (DSOs) are equipped with flight recording capabilities, allowing you to monitor and capture real-time vehicle data during dynamic testing—such as a road test. This feature allows you to observe intermittent faults that may not appear in a static workshop environment, ensuring more precise diagnostics.

Advanced Techniques

How Flight Recording Works

- During a road test, the oscilloscope continuously records waveforms, capturing live sensor signals, actuator responses, and network communications.
- You can set up trigger points or alarm conditions—these act as automatic markers that flag unusual events, such as voltage drops, signal distortions, or erratic component behaviour.
- Once the test concludes, the stored data can be reviewed frame by frame, allowing a deeper post-analysis in the workshop without the pressure of live driving conditions.

Key Benefits of Flight Recording

- Diagnosing Intermittent Faults: Captures sporadic electrical issues that might not surface in a static test.
- Correlation with Driving Conditions: Helps link waveform abnormalities with real-world scenarios like acceleration, braking, or rough road conditions.
- Efficient Post-Test Analysis: Allows you to replay, zoom in, and compare waveform behaviours with known good reference signals, improving diagnostic accuracy.

By leveraging flight recording, you gain invaluable insights into complex vehicle behaviour, ensuring precise fault identification that would otherwise be difficult to replicate in a stationary test.

Real-time monitoring

Use

Step 1: Begin capturing waveforms as the vehicle is operated or driven under various conditions.

Step 2: Use an assistant or flight record where possible and monitor the signals in real-time to detect any anomalies.

Step 3: Record waveforms during different phases such as idle, acceleration, cruising, and braking.

Advanced Techniques

Interpretation

Use

Step 1	Analyse the captured waveforms to identify patterns and inconsistencies.
Step 2	Compare the signals against reference waveforms (known good) to spot deviations.
Step 3	Use waveform analysis techniques to diagnose faults including those that may arise from interconnected systems.
Step 4	Refer to real-world case studies or internet forums for a detailed breakdown of similar diagnostic challenges.

Documentation

Use

Step 1	Document the findings meticulously.
Step 2	Record all observations and the corresponding waveforms.
Step 3	Note down common missteps and troubleshooting tips for future reference.

Best practices for learning oscilloscope diagnostics

Educational Insights

To master the interpretation of dynamic waveforms, consider the following tips:

- Engage in continuous learning and training on oscilloscope usage.
- Participate in workshops and hands-on sessions.
- Research various case studies to understand different diagnostic scenarios.

Avoiding Pitfalls

Common missteps in setup and analysis can hinder accurate diagnostics. It's essential to:

- Ensure proper calibration of the oscilloscope.
- Avoid incorrect probe placement which can lead to erroneous readings.
- Double-check settings before capturing waveforms.

Advanced Techniques

Measurements

Some oscilloscopes offer various preprogrammed measurement options. Measurement functions extract key information from a capture and present the data in a table for analysis. This data can be captured live or logged to display information collected over time.

To access these functions, enter the measurements menu and select the channel and type of measurement needed from a list. Measurements can be applied to the entire trace or focused on a specific area between rulers.

Table 6.2 provides an overview of the types of measurements that may be available, their applications, and their significance.

Measurement parameters will be preprogrammed by the oscilloscope manufacturer and shouldn't require adjustment unless specific diagnostic conditions need to be met.

Table 6.2 Waveform measurement styles

Measurement	Description	Usage	Importance
Time	Refers to the duration of a waveform cycle.	Used to determine the period and frequency of a signal.	Critical for understanding signal timing and identifying frequency-related issues.
Frequency	Number of waveform cycles per second.	Measured to determine the periodicity of signals.	Vital for diagnosing components reliant on specific frequency ranges.
Cycle time	The time taken for one complete waveform cycle.	Used to calculate the period of the signal.	Essential for accurate timing analysis.
Negative duty cycle	Percentage of time the signal is in the low state.	Measured to understand signal behaviour during its low phase.	Important for assessing signals with varying duty cycles.
Positive duty cycle	Percentage of time the signal is in the high state.	Measured to comprehend signal behaviour during its high phase.	Important for assessing signals with varying duty cycles.
Edge count	Total number of transitions between high and low states.	Used to count signal events over a specified period.	Useful for tracking activity in digital signals.
Rising edge count	Number of transitions from low to high.	Measured to count upward transitions in a signal.	Important for detecting positive transitions in pulse signals.
Falling edge count	Number of transitions from high to low.	Measured to count downward transitions in a signal.	Essential for observing negative transitions in pulse signals.
High pulse width	Duration of the high state within a pulse.	Measured to understand the length of high pulses in a signal.	Critical for pulse width modulation analysis. Important for understanding the signals off period or on-time if pull-up circuit.

Advanced Techniques

Table 6.2 Waveform measurement styles

Measurement	Description	Usage	Importance
Low pulse width	Duration of the low state within a pulse.	Measured to assess the length of low pulses in a signal.	Important for understanding the signals off period or on-time if pull-down circuit.
Rise time	Time taken for the signal to transition from low to high.	Measured to analyse the speed of upward transitions.	Essential for evaluating fast signal changes.
Fall time	Time taken for the signal to transition from high to low.	Measured to analyse the speed of downward transitions.	Critical for understanding the decay rate of signals.
Rising rate	Rate of signal increase during the rise time.	Used to measure the steepness of upward transitions.	Important for detecting rapid signal changes.
Falling rate	Rate of signal decrease during the fall time.	Used to measure the steepness of downward transitions.	Essential for observing signal decay characteristics.
Amplitude	Peak value of signal voltage.	Measured to determine the maximum voltage level.	Critical for assessing signal magnitude and strength.
Minimum	Lowest point of signal voltage.	Measured to identify the minimum voltage level.	Important for evaluating signal baselines.
Base	Reference level of the signal.	Used as a baseline for other measurements.	Essential for accurate signal assessment.
Negative overshoot	Extent of signal drop below the base level.	Measured to understand negative excursions.	Important for analysing signal distortions.
Maximum	Highest point of signal voltage.	Measured to identify the maximum peak.	Critical for assessing signal peaks.
Top	Uppermost level of the signal.	Used for peak analysis.	Important for understanding signal boundaries.
Positive overshoot	Extent of signal rise above the top level.	Measured to analyse positive excursions.	Essential for evaluating signal distortions.
Peak to peak	Difference between the minimum and maximum signal levels.	Used to measure signal range.	Critical for understanding overall signal amplitude.
Mean	Average value of the signal.	Measured to determine signal average.	Important for overall signal analysis.
RMS	Root Mean Square value of the signal.	Used to measure signal power.	Critical for evaluating signal energy.
RMS ripple	Variations in RMS value.	Measured to analyse signal consistency.	Important for assessing signal stability.
Power	Rate at which electrical energy is transferred.	Measured to determine signal power.	Critical for evaluating energy consumption.
True power	Actual power consumed by the load.	Measured to calculate real energy usage.	Essential for accurate power analysis.
Apparent power	Total power in the circuit.	Used to measure overall power.	Important for understanding total energy transfer.
Reactive power	Power stored and released by the load.	Measured to assess unused power.	Critical for power factor analysis.
Power factor	Ratio of true power to apparent power.	Used to measure efficiency.	Essential for optimising power usage.
DC power	Power in direct current circuits.	Measured to determine DC energy usage.	Important for DC system analysis.

Advanced Techniques

Table 6.2 Waveform measurement styles

Measurement	Description	Usage	Importance
Crest factor	Ratio of peak value to RMS value.	Measured to analyse signal peaks.	Critical for evaluating signal peaks.
Area at AC	Accumulated area under the AC signal.	Used for energy analysis.	Important for understanding overall AC energy.
Area positive AC	Accumulated area under the positive AC signal.	Used to measure positive AC energy.	Essential for positive signal analysis.
Area at negative AC	Accumulated area under the negative AC signal.	Used to measure negative AC energy.	Important for negative signal analysis.
Absolute area at AC	Total area under the AC signal.	Used for comprehensive energy analysis.	Critical for understanding total AC energy.
Area at DC	Accumulated area under the DC signal.	Used to measure DC energy.	Important for DC signal analysis.
Positive area at DC	Accumulated area under the positive DC signal.	Used to measure positive DC energy.	Essential for positive DC signal analysis.
Negative area at DC	Accumulated area under the negative DC signal.	Used to measure negative DC energy.	Important for negative DC signal analysis.
Absolute area at DC	Total area under the DC signal.	Used for comprehensive DC energy analysis.	Critical for understanding total DC energy.
Multi-Channel	Analysis of multiple signals simultaneously.	Used to compare and correlate signals.	Important for complex system diagnostics.
Phase	Difference in timing between two signals.	Measured to assess signal synchrony.	Critical for timing analysis.
Delay	Time difference between signal events.	Used to measure event timing.	Important for synchrony assessment.

Measurements provide quantitative data about a waveform that may not be visible when viewing a capture. While some measurement styles may be unnecessary, certain ones can be essential for specific types of diagnosis.

Measurement data is useful in situations where the timing of components or systems is important. For instance, phase can be used to determine the correct camshaft to crankshaft correlation when engine timing issues are suspected. This can be particularly beneficial when assessing a stretched cam-chain. Comparing known good values can provide an effective initial 'first look' diagnosis.

Deep measure

Although considered a very advanced and time consuming function, deep measure is a powerful feature of oscilloscopes that allows you to extract a wealth of information from captured waveforms. By utilising deep measure functions, you can analyse intricate details and subtle variations within a signal, which are often crucial for accurate diagnostic and fault isolation in complex automotive systems.

Advanced Techniques

What is deep measure?

Deep measure involves advanced measurement capabilities that extend beyond basic waveform observation. These functions include statistical analysis, automated measurement of waveform parameters, and long-term data collection. Deep measure is valuable for providing detailed analysis, allowing for the identification of issues that may not be evident through simple visual inspection.

How deep measure can be used

Deep measure functions can be used in multiple scenarios for diagnostics:

- Statistical Analysis: By measuring and analysing a large number of waveform cycles, you can identify patterns and anomalies that indicate underlying issues.
- Automated Measurements: Automated tools can quickly calculate parameters such as frequency, amplitude, rise time, and more, saving time and reducing human error.
- Long-term Data Collection: Capturing data over extended periods allows for the observation of intermittent faults and gradual changes in signal behaviour.

Masks

Masks are a feature that define acceptable boundaries for waveform signals. By creating a visual mask around a waveform, you can see if a captured signal falls inside or outside specified limits. If a signal breaches the mask boundaries, it leaves an imprint that may require further investigation.

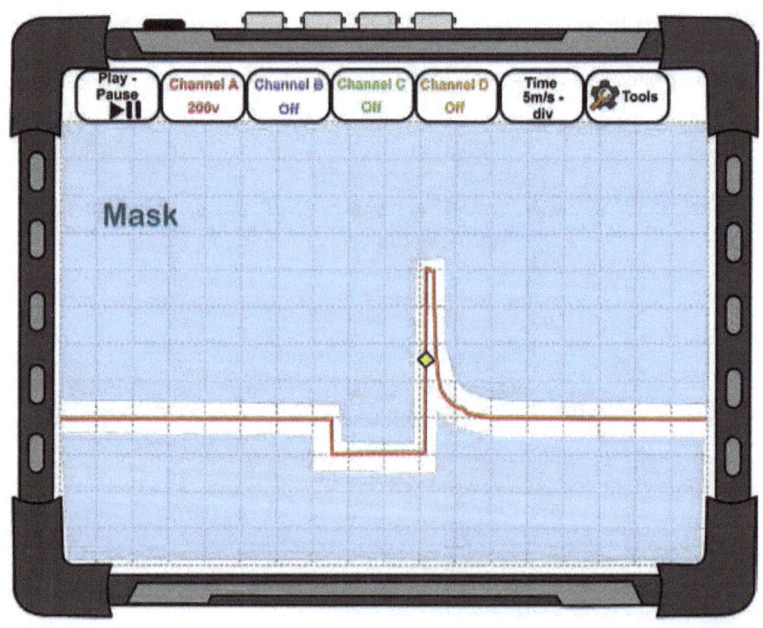

Figure 6.3 Waveform mask

Using waveform masks on an automotive oscilloscope is like marking lanes on a racetrack to ensure every car stays within the expected path. Just as race officials set boundaries to detect if a car drifts off course, a waveform mask allows you to determine if an electrical signal deviates from acceptable limits—helping pinpoint faults quickly and accurately.

A waveform mask is a predefined boundary that overlays an expected waveform shape on the oscilloscope screen. Any deviation outside this mask triggers an alert, indicating a possible problem. This method helps detect erratic voltage fluctuations, timing inconsistencies, or intermittent faults without requiring constant monitoring.

Advanced Techniques

Setup and Use

Step 1
- Capture a Reference Waveform: Set up the oscilloscope and begin by capturing a known good waveform that represents the expected signal under normal operating conditions.

Step 2
- Define Mask Boundaries: Using the oscilloscope's mask setup function, create the boundaries around the reference waveform. These boundaries represent the upper and lower limits of acceptable signal variation. Masks can often be automatically generated from the source image.

Step 3
- Apply the Mask: Once the mask is defined, enable the mask testing feature. The oscilloscope will now compare incoming waveforms against the mask.

Step 4
- Adjust Sensitivity: Fine-tune the mask boundaries to ensure that they accurately reflect your acceptable range of signal variations. This step may involve adjusting the tolerance levels for specific parameters such as amplitude and frequency.

Step 5
- Test the Component or Circuit: Operate the system and monitor the image output for anomalies.

Step 6
- Set Actions: If necessary, use the oscilloscope action functions to monitor the image output and help identify signal irregularities. This could involve stopping/restarting the capture, saving, starting an application, exporting serial data or sounding an alarm.

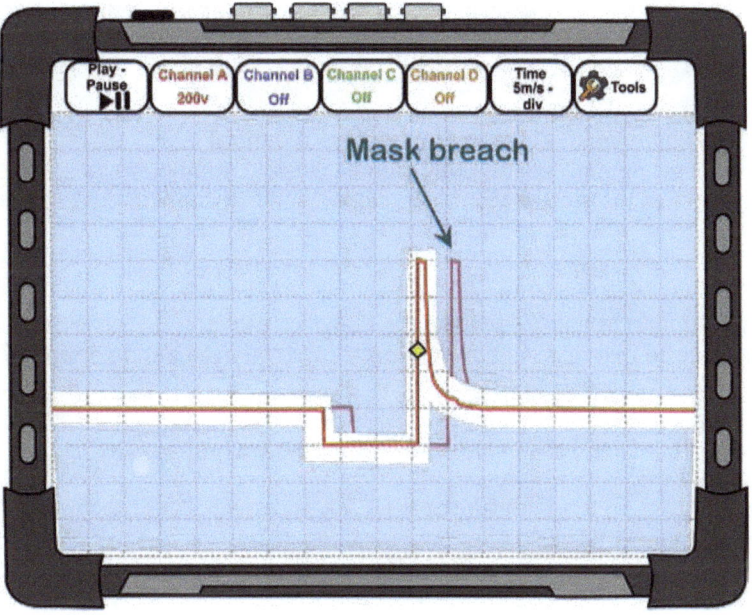

Figure 6.4 Mask breach

Why masks are useful

Masks are incredibly useful for several reasons:

- Quick Fault Detection: Masks enable rapid identification of signal anomalies, saving valuable diagnostic time.
- Consistency: Masks provide a consistent and objective method for evaluating waveforms, reducing the potential for human error.
- Documentation: Masks can be saved and reused, providing a reliable method for documenting and comparing waveforms over time.
- Automation: Mask testing can be automated, allowing for continuous monitoring of signal integrity without constant user intervention.

Example scenarios

Masks can be used in various diagnostic scenarios, including:

- Engine Diagnostics: Masks can be used to monitor signals from sensors such as crankshaft position sensors or camshaft position sensors. Any deviation from the expected waveform can indicate issues such as timing chain stretch or sensor malfunction.
- Electrical System Testing: When diagnosing issues in the vehicle's electrical system, masks can help identify irregularities in voltage signals from components such as the alternator, battery, or electronic control units (ECUs).
- Communication Networks: In complex automotive networks such as CAN or LIN BUS systems, masks can be used to detect errors in data transmission. This is crucial for identifying intermittent faults or communication breakdowns.
- Intermittent Faults: For faults that occur sporadically, masks allow for long-term monitoring of signals. When an anomaly occurs, the oscilloscope can log the event, aiding in the diagnosis of elusive issues.

Actions

With a lot of automotive oscilloscopes, the 'Action Menu' plays a central role in enhancing your workflow efficiency. Whether you're capturing high-speed CAN signals or slow analogue signals from a throttle position sensor, understanding how to use the Action Menu can significantly streamline diagnostics and documentation.

What is the action menu?

The Action Menu is a feature found in many oscilloscope software platforms. It provides quick access to commonly used tasks that can be executed either automatically or at the user's command, depending on how the oscilloscope is set up.

In simple terms, actions are predefined tasks the oscilloscope can perform either:

When a specific event occurs (e.g., a trigger fires), or
Manually, when selected by the technician.

Common actions available

While options may vary between scope brands, **Table 6.3** provides some examples of standard actions typically found in automotive oscilloscopes.

Advanced Techniques

Table 6.3 Standard actions typically found in automotive oscilloscopes

Action	Description
Stop capture	Automatically stops or pauses a waveform capture when the trigger condition is met (e.g., a voltage spike on a pedal position sensor).
Save on trigger	Automatically saves a waveform file when the trigger condition is met (e.g., a voltage drop on a crank sensor).
Sound alarm	Plays a sound when a trigger occurs — useful when monitoring intermittent faults or leaving the scope unattended.
Run external script or program	Some advanced scopes can run custom applications, scripts or notify diagnostic tools when a specific signal condition is detected.
Take screenshot	Automatically captures a screenshot of the waveform display. Ideal for documentation or customer reporting.
Export serial data	Decodes and exports the CAN BUS data when a measurement failure limit is exceeded.
Email notification	Sends an alert or image by email — useful in remote diagnostic environments (available in networked scopes).

Accessing the action menu

Typically, the Action Menu can be accessed through:

- The main toolbar (labelled 'Actions')
- A right-click context menu on the waveform display area
- Or under the trigger configuration settings

Some software also allows for per-channel action assignment, letting you set up different reactions for different signals (e.g., CAN_H vs Crankshaft signal).

Using actions in real-world diagnostics

Example 1: intermittent crankshaft sensor dropout

You suspect a crankshaft position sensor is intermittently losing signal. Rather than watching the screen constantly, you can set an Action to 'Save on Trigger' when the signal drops below 0.5V. Now, each time the fault occurs, the oscilloscope logs it without your input — capturing valuable evidence.

Example 2: capture during road test

While test driving, you set the oscilloscope to capture and save whenever a misfire pattern is detected on the ignition signal. The Action Menu lets you focus on driving while the scope takes care of catching the fault.

Advanced Techniques

Best practices

- Always test your action setup before relying on it during a diagnostic session.
- Combine actions with triggers for maximum precision — e.g., "Save when trigger fires AND signal < threshold."
- Label saved files and screenshots for easy reference.

The Action Menu transforms your oscilloscope from a passive measuring tool into an active diagnostic assistant. By automating routine tasks such as saving waveforms or sounding alarms, you can focus more on analysis and less on manual data collection.

Reference Waveforms

Reference waveforms are standard patterns or signals used to compare live data during diagnostics. They show the ideal or expected behaviour of a component or system under normal conditions. By using reference waveforms, you can quickly identify any deviations in measured signals, making it easier to diagnose potential issues accurately.

Creating reference waveforms

Creating reference waveforms involves capturing and storing signals from a properly functioning component or system.

Setup

Step 1
- Select the Component: Choose the component or system you wish to create a reference for, such as a sensor, actuator, or communication network.

Step 2
- Setup the Oscilloscope: Properly configure the oscilloscope to capture the signal. This includes setting the correct voltage range, time base, and any triggering conditions.

Step 3
- Record the Signal: Capture the waveform from the component during normal operation. Ensure that the signal is stable and repeatable.

Step 4
- Analyse the Signal: Examine the captured waveform for consistency and accuracy. Verify that it represents the ideal operating conditions of the component.

Step 5
- Save the Waveform: Store the waveform in the oscilloscope's memory or export it to external storage. Label it clearly with relevant details such as the component name, operating conditions, and date.

Advanced Techniques

Reference waveforms provide several advantages in automotive diagnostics:

- Consistency: They offer a standardised method for evaluating signals, reducing variability and human error.
- Documentation: Saved reference waveforms act as reliable documentation for comparing signals over time, aiding in historical analysis.
- Efficiency: You can quickly compare live signals against reference waveforms, speeding up the diagnostic process.
- Automation: Reference waveforms enable automated testing and monitoring, allowing continuous oversight without constant user intervention.

Saving reference waveforms for future use

Ensure reference waveforms are readily available for future diagnostics.

Setup

Step 1
- Label and Organise: Clearly label each waveform with relevant details and organise them systematically in the oscilloscope or external storage.

Step 2
- Backup: Regularly back up waveforms to prevent data loss and ensure they are accessible when needed.

Step 3
- Update: Periodically review and update reference waveforms to reflect any changes in component design or operating conditions.

Step 4
- Share: Share reference waveforms with colleagues or store them in a centralised database for collaborative diagnostics.

How to get the most from reference waveforms:

- Build a library of known good signals.
- Align time and voltage scales for accurate comparisons.
- Use overlay features for real-time diagnosis.
- Verify variations across different operating conditions.

Advanced Techniques

Real-world diagnostic examples

Table 6.4 provides some example scenarios where reference waveforms could be used:

Table 6.4 Real-world diagnostic examples

Example	Scenario
Engine diagnostics	Reference waveforms can be used to monitor signals from sensors such as crankshaft position sensors or camshaft position sensors. Any deviation from the expected waveform could indicate issues such as timing chain stretch or sensor malfunction.
Electrical system testing	When diagnosing issues in the vehicle's electrical system, reference waveforms help identify irregularities in voltage signals from components such as the alternator, starter, or electronic control units (ECUs).
Communication networks	In complex automotive networks such as CAN or LIN BUS systems, reference waveforms help detect errors in data transmission. This is crucial for identifying intermittent faults or communication breakdowns.
Intermittent faults	For faults that occur sporadically, reference waveforms allow for long-term monitoring of signals. When an anomaly occurs, the oscilloscope can log the event, aiding in the diagnosis of elusive issues.

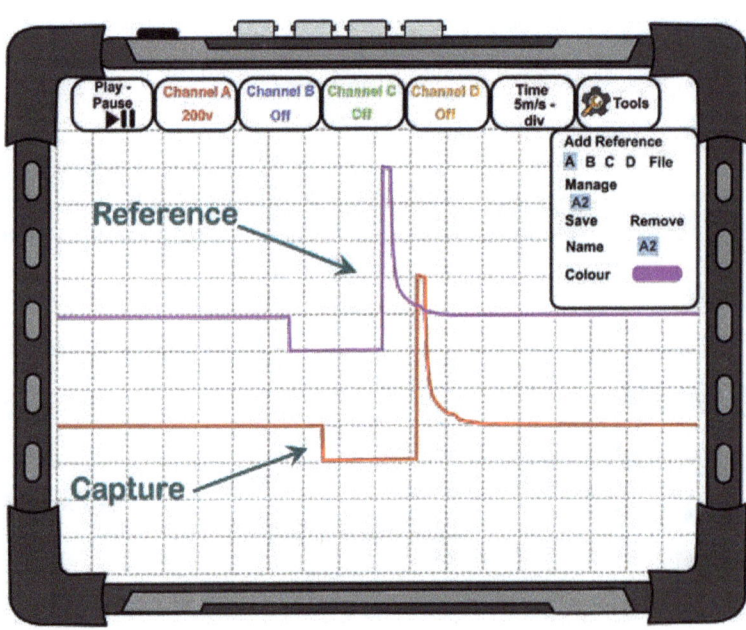

Figure 6.5 Reference waveforms

Advanced Techniques

Maths Channels

Maths channels are a powerful feature which allows you to conduct an arithmetical calculation on a waveform capture and display the result as a physical plot on the oscilloscope screen. It also lets you compare two or more waveforms and enables the comparison of multiple signals with the extraction of meaningful information that can help guide your diagnostics. These operations can include simple calculations, such as addition and subtraction, as well as more complex functions like differentiation and integration.

Setting up maths channels

Setup

Step 1
- Select the channel(s) that you want to use for the maths operation. Typically, these are the channels where the waveforms you want to compare or analyse are displayed.

Step 2
- Access the maths channel menu on the oscilloscope. This menu is usually found in the advanced features section.

Step 3
- Choose the type of mathematical operation you want to perform. Oscilloscopes may offer a range of preset calculations including addition, subtraction, multiplication, division, differentiation, and integration.

Step 4
- Link the waveforms/channels to the equations or mathematical functions that will provide the desired output information. This can be done through the oscilloscope's interface, which may include a keypad or touchscreen.

Step 5
- Configure the display settings for the maths channel. You can often choose to view the results as numerical values or physical plots on the oscilloscope screen.

Step 6
- Run the oscilloscope to capture and analyse the data using the configured maths channels.

ANALOGY

Using maths channels in an automotive oscilloscope is like having a translator for a complex conversation—it takes raw waveform data and converts it into meaningful insights that would otherwise be difficult to interpret.

Imagine driving a car without a fuel efficiency gauge. You can see how fast you're going, and you know how much fuel is in the tank, but without additional calculations, you don't have a precise measure of miles per gallon (MPG).

Advanced Techniques

Applications of maths channels

Maths channels can provide a wealth of information that is crucial and informative for automotive diagnostics.

Table 6.5 shows some key applications:

Table 6.5 Applications of maths channels

Maths channel equation	How it's useful	Real-world examples (case studies)
A - B (signal subtraction) **Compares the difference between two signals.**	Helps you see the potential difference between two channels. A powerful and precise tool for comparing related signals — ideal for catching subtle faults that might be missed on standard waveforms.	**1. Crankshaft vs Camshaft Sensor – Timing Correlation** • Application: Diagnosing cam/crank timing issues (e.g., P0016, P0340). • Why Use A-B: Subtracting the camshaft signal (B) from the crankshaft signal (A) highlights any shift in phase between them. • How It Helps: Reveals whether the cam and crank signals are properly synchronised. A phase shift or misalignment in the A-B waveform could indicate a stretched timing chain, slipped timing belt, or incorrect mechanical timing.
		2. Throttle Position Sensor (TPS) Redundant Tracks – Signal Comparison • Application: Many TPS sensors have two tracks (e.g., TPS1 and TPS2) that move in opposite or mirrored directions. • Why Use A-B: Subtracting one TPS signal from the other should result in a steady, predictable value. • How It Helps: Any irregularity or spikes in the A-B result indicate a faulty sensor, worn tracks, or poor internal alignment, especially useful for drive-by-wire throttle systems.
		3. Injector Current vs Voltage Drop – Electrical Integrity Check • Application: Diagnosing faulty injector wiring, poor grounds, or internal resistance. • Why Use A-B: Subtracting the injector voltage (B) from the current trace (A) reveals how voltage and current correlate during activation. • How It Helps: If the A-B result shows inconsistent drops or delays, it may point to resistance in the power/ground circuits, poor connectors, or failing injector coils.
		4. Wheel Speed Sensor Comparison – ABS Fault Diagnosis • Application: Diagnosing intermittent ABS light or false activation of ABS. • Why Use A-B: Subtract one wheel speed sensor (e.g., front-left) from another (e.g., front-right) while driving. • How It Helps: A-B shows discrepancies in output that may not be visible when looking at each signal separately. Differences in the A-B waveform can indicate tone ring damage, sensor dropout, or misalignment.

Advanced Techniques

Table 6.5 Applications of maths channels

Maths channel equation	How it's useful	Real-world examples (case studies)
A + B (signal addition) Adds two signals together into one trace.	Lets you see the total activity or overlap of two systems. Can simplify complex signal relationships, helping you confirm event coordination, network integrity, and system balance.	**1. Turbocharger Control: Wastegate Command + Position Feedback – Turbo Response Analysis** • Application: Diagnosing underboost, overboost, or turbo lag complaints (e.g., P0299). • How A+B Helps: Add the wastegate actuator control signal (A) and the position sensor feedback (B). • What It Reveals: Helps compare command vs actual position in a simplified trace. Any discrepancy in the summed signal can point to sticking actuators, vacuum leaks, or controller faults delaying boost control.
		2. CAN High + CAN Low – CAN BUS Health Check • Application: Diagnosing intermittent communication issues or network-related faults (e.g., U-codes). • Why Use A+B: CAN High (A) and CAN Low (B) are mirror images around 2.5V. When working properly, A + B should result in a flat line (around 5V). • How It Helps: If the sum waveform fluctuates or has noise/spikes, it indicates network corruption, grounding issues, or termination resistor problems. It simplifies CAN signal validation without needing to interpret each line individually.
		3. EVAP System Monitoring: Purge Valve Command + Fuel Tank Pressure Sensor – Leak or Flow Verification • Application: Diagnosing EVAP system faults like P0441 (incorrect purge flow) or P0455 (large leak). • How A+B Helps: Add the purge solenoid duty cycle signal (A) and the tank pressure sensor voltage (B). • What It Reveals: Creates a combined view showing when purge occurs and how the tank responds. If there's command but no change in pressure, it could indicate a blocked line, stuck valve, or leak in the system.
		4. Transmission Speed Sensors: Input Shaft Speed (ISS) + Output Shaft Speed (OSS) – Gear Ratio Validation • Application: Diagnosing harsh shifting, slip, or gear ratio error codes (e.g., P0730–P0736). • How A+B Helps: Add the Input Shaft Speed signal (A) and the Output Shaft Speed signal (B). • What It Reveals: A stable, predictable sum across gears indicates normal ratio transitions. Deviations or sudden changes in the combined waveform can reveal clutch pack slippage, internal wear, or speed sensor signal dropout.

Advanced Techniques

Table 6.5 Applications of maths channels

Maths channel equation	How it's useful	Real-world examples (case studies)
A ÷ B (division) **Compares how one signal behaves relative to another, especially in sensor relationships, ratios, or signal validation.**	Compare behaviours, calculate dynamic ratios, and normalise signal differences, offering a clearer understanding of how systems interact in real time. Can display operating resistance if voltage divided by current (Ohms Law).	1. Camshaft RPM ÷ Crankshaft RPM – Cam/Crank Timing Accuracy • Application: Diagnosing timing chain stretch or variable valve timing faults (e.g., P0016). • How A ÷ B Helps: Cam RPM (A) ÷ Crank RPM (B) gives a mechanical ratio that should remain stable. • What It Reveals: Deviations in this ratio suggest timing chain wear, slippage, or phaser faults.
		2. Throttle Position Sensor ÷ Pedal Position Sensor – Drive-by-Wire Consistency Check • Application: Diagnosing throttle control issues or hesitation (e.g., P2101, P2135). • Why Use A ÷ B: Divide throttle plate position (A) by accelerator pedal position (B) to assess throttle response to driver input. • How It Helps: If the ratio isn't steady or as expected (e.g., too slow to rise), it could point to a lazy throttle motor, sensor mismatch, or software control problem. Helps verify electronic throttle integrity.
		3. Fuel Rail Pressure Sensor ÷ Command Signal – Pressure Control Feedback • Application: Diagnosing rail pressure control faults or response lag. • How A ÷ B Helps: Divides the actual pressure signal (A) by the command signal from the Powertrain Control Module PCM (B). • What It Reveals: Helps confirm whether fuel pressure responds proportionally to control commands. Lagging or unstable ratio suggests pump weakness, leaking injectors, or pressure regulator faults.
		4. Battery Voltage ÷ Charging Current – Charging System Load Check • Application: Diagnosing undercharging or overcharging issues. • Why Use A ÷ B: Divide charging voltage (A) by charging current (B) to evaluate volts per amp. • How It Helps: A very high or low ratio could indicate high resistance in the charging circuit, a weak alternator, or a poor battery accepting charge. Helps validate alternator efficiency under load.

Advanced Techniques

Table 6.5 Applications of maths channels

Maths channel equation	How it's useful	Real-world examples (case studies)
A × B (multiplication) **Multiplies two signals together.**	Commonly used to evaluate power (voltage × current) or other interactions where signal amplitude changes matter. The A×B maths channel is a powerful diagnostic tool when you need to assess electrical load, component efficiency, or real-world performance under demand — especially when voltage and current alone don't tell the full story.	1. Fuel Pump Power (Voltage × Current) – Checking Fuel Pump Load • Application: Diagnosing fuel delivery issues (e.g., long crank, low pressure, engine stall). • Why Use A × B: Multiply fuel pump voltage (A) by current draw (B) to calculate real-time power (Watts). • How It Helps: Helps identify an overloaded or underperforming fuel pump. A healthy pump draws consistent power; dropping or spiking values may point to worn brushes, restricted filters, or impending failure. 2. Electric Cooling Fan Voltage × Current – Load and Response Diagnosis • Application: Overheating complaints or fan not activating at proper temperatures. • How A × B Helps: Multiply fan supply voltage (A) and fan current (B). • What It Reveals: Tracks fan power draw. A declining power curve can signal brush wear, bearing resistance, or voltage drop in wiring/relays. 3. Glow Plug Power – Diesel Cold Start Performance • Application: Diagnosing poor cold starts or extended cranking in diesel engines. • Why Use A × B: Multiply glow plug voltage (A) by current (B) to check actual power per plug. • How It Helps: Glow plugs should produce consistent power across all cylinders. If one shows low wattage, it may be failing, open circuit, or have a control issue, even if resistance appears normal at rest. 4. Heated Oxygen Sensor Power – Sensor Readiness Check • Application: Diagnosing delayed closed-loop operation or failed emissions readiness. • Why Use A × B: Multiply heater voltage (A) by heater current (B) for the O2 sensor. • How It Helps: Helps verify the O2 sensor reaches operating temperature quickly. Low power = slow warm-up = delayed closed-loop. Can reveal open heater elements, wiring faults, or PCM control issues.

Advanced Techniques

Table 6.5 Applications of maths channels

Maths channel equation	How it's useful	Real-world examples (case studies)
Low Pass Filter (A) **Smooths out high-frequency 'noise' in the signal.**	Makes it easier to see slow changes in signals by filtering out unwanted spikes. Useful for cleaning up noisy signals and focusing on slow-moving or average trends. Low Pass Filter A helps: Smooth noisy or pulsed signals. Reveal underlying patterns. Isolate slow or gradual faults. Enhance sensor diagnostics.	1. Throttle Position Sensor (TPS) – Diagnosing Dead Spots or Dropouts • Application: Hesitation on acceleration, inconsistent throttle response, or surging. • Why Use Low Pass Filter A: TPS signals can be noisy due to vibration or internal wear. • How It Helps: The filter smooths out high-frequency noise, making it easier to see gradual transitions or flat spots in the signal. Helps identify dead zones, dropouts, or glitches that may be missed in a raw waveform. 2. Fuel Tank Pressure Sensor – EVAP System Testing • Application: Diagnosing EVAP system codes (e.g., P0440, P0455). • Why Use Low Pass Filter A: Pressure readings can have small fluctuations due to pump pulsing or sensor chatter. • How It Helps: The low pass filter removes those quick spikes so you can see the true pressure trend as the system seals and builds vacuum. This helps detect leaks, faulty purge valves, or sensor issues more clearly. 3. Manifold Absolute Pressure (MAP) Sensor – Idle and Load Evaluation • Application: Poor idle, stalling, or lack of power complaints. • Why Use Low Pass Filter A: MAP signals fluctuate rapidly with engine vacuum pulses, especially at idle. • How It Helps: Filtering the signal provides a clear average pressure level, making it easier to see trends over time and detect issues like low vacuum, leaks, or slow-responding sensors. 4. Accelerator Pedal Position Sensor – Drive-by-Wire Input Analysis • Application: Hesitation, throttle lag, or unresponsive acceleration. • Why Use Low Pass Filter A: Driver foot movement and road vibration can introduce rapid signal changes. • How It Helps: Filtering the APP signal reveals the intentional pedal movements and makes it easier to compare with throttle body response. This helps identify lag, mismatch, or sensor faults in electronic throttle control systems.

Advanced Techniques

Table 6.5 Applications of maths channels

Maths channel equation	How it's useful	Real-world examples (case studies)
Frequency(A) or RPM from A **Calculates how fast a repetitive signal occurs.** **A powerful way to convert repeating signal patterns—such as sensor pulses—into understandable data like RPM or cycles per second.**	Useful for identifying mechanical faults, sensor dropouts, or signal inconsistencies that affect drivability. Converts sensor waveforms into usable speed or RPM data. Using Frequency(A) or RPM(A) helps: Confirm rotating component activity in non-starts. Detects signal dropouts not visible in waveform shape alone. Tracks dynamic timing relationships. Validates sensor input for proper ECU operation.	1. Crankshaft Position Sensor – RPM Monitoring During Non-Start • Application: Engine cranks but doesn't start (no-spark, no-fuel conditions). • Why Use Frequency(A) or RPM(A): Converts the crank sensor's waveform into a real-time RPM signal. • How It Helps: If no or slow RPM is detected during cranking, it could point to a faulty crank sensor, wiring issue, or ECU not receiving signal. Quickly confirms whether the engine's rotational signal is reaching the Powertrain Control Module PCM. 2. Wheel Speed Sensor (WSS) – ABS Fault Diagnosis • Application: Intermittent ABS light or traction control activation. • Why Use Frequency(A): Converts wheel speed pulses into Hz (cycles/second) to show wheel rotation speed. • How It Helps: Comparing frequency from multiple WSS channels reveals mismatched signals, which can point to sensor dropouts, damaged tone rings, or excessive air gap causing false ABS triggers. 3. Camshaft Sensor Frequency – Timing Chain Health Check • Application: Rattling noise, timing-related DTCs (e.g., P0016, P0341). • Why Use Frequency(A): Tracks the camshaft signal's cycle rate to assess cam movement relative to crankshaft. • How It Helps: In combination with crank frequency, a change in cam signal frequency can reveal timing chain stretch, slip, or improper valve timing, especially at startup when slack is most likely. 4. Diesel Injector Control Signal – High-Pressure Pump RPM Verification Application: Hard start, poor performance, or fuel rail pressure DTCs. Why Use Frequency(A): On some diesels, the high-pressure fuel pump or fuel metering solenoid is pulse-controlled. How It Helps: Monitoring injector control or pump actuator frequency can confirm pump speed, and whether the PCM is commanding fuel delivery. A flat or missing frequency shows a potential driver or supply issue.

Advanced Techniques

Table 6.5 Applications of maths channels

Maths channel equation	How it's useful	Real-world examples (case studies)
Invert A **Inverts the waveform (turns it upside-down).**	Helpful when you're trying to compare signals that are mirror images, work in opposition, or when you need to align events visually for analysis. Using Invert A helps you: Visually align mirror-image signals. Simplify comparison between opposing or inverse channels. Detect subtle phase shifts or signal integrity problems. Clarify polarity or edge-triggered events.	**1. CAN BUS Diagnostics – Comparing CAN High and CAN Low** • Application: Diagnosing intermittent communication faults or U-codes. • Why Use Invert A: CAN High and CAN Low are mirror images of each other. If Channel A is CAN Low, inverting it makes it align visually with CAN High on Channel B. • How It Helps: Makes it easy to visually confirm proper signalling, symmetry, and timing. Misalignment could indicate wiring issues, interference, or a missing termination resistor. **2. Crankshaft vs. Camshaft Signal Alignment – Phase Comparison (digital)** • Application: Diagnosing cam/crank correlation faults (e.g., P0016, P0340). • Why Use Invert A: If the cam and crank signals are out of phase but hard to interpret, inverting one of them can help visually align key transitions. • How It Helps: Makes it easier to identify a timing shift or delayed synchronisation, helping spot issues like timing chain stretch, misaligned timing marks, or slipped reluctors. **3. Accelerator Pedal Position Sensor Redundancy Check** • Application: Drive-by-wire throttle issues, hesitation, or pedal DTCs (e.g., P2122, P2135). • Why Use Invert A: Many APP sensors use two opposing signals—one increases while the other decreases. Invert one of the signals to make both rise together. • How It Helps: Simplifies comparison and visual matching. Any mismatch between the inverted and non-inverted signals may indicate sensor wear, misalignment, or wiring faults. **4. Hall Effect Sensors – Polarity Alignment Across Channels** • Application: Diagnosing sensor input issues or missing signal edges (e.g., ABS, VSS, crank/cam sensors). • Why Use Invert A: Some Hall sensors are wired with reversed polarity or different pull-up configurations. • How It Helps: Inverting the signal lets you match rising and falling edges to other channels, helping verify sensor synchronisation, missing pulses, or wrong voltage swing direction.

Advanced Techniques

Maths channels can be used to easily display the percentage efficiency, power loss and operating resistance of electrical components.

Calculation	Oscilloscope Setup	Description	Math Channel Equation	Units
% Efficiency	**Channel A**: System voltage (e.g., battery feed) **Channel B**: Volt drop across component.	Calculates what percentage of system voltage is successfully delivered to the component (not lost across it).	$((A - B) \div A) \times 100$	% (percent)
Power Loss	**Channel B**: Volt drop across component **Channel C**: Current through component (via current clamp).	Calculates real-time power being wasted or dissipated across a component as heat or loss.	$B \times C$	Watts (W)
Resistance	**Channel B**: Volt drop across component **Channel C**: Current through component (via current clamp).	Calculates the effective resistance of the component under load. Useful for motor windings, connectors, wiring.	$B \div C$	Ohms (Ω)

Advanced maths channel functions could include:
- Differentiation: dV/dt (Analysing the rate of voltage change over time to detect rapid fluctuations).
- Integration: ∫V dt (Calculating the total accumulated voltage over time for energy analysis).
- Phase Calculation: φ = arctan(V2/V1) (Determining the phase angle between two sinusoidal signals).

Error Analysis in Complex Networks

In-vehicle **networks** are integral to the design and structure of most modern vehicles. They enable **multiplexing**, which is essential for the operation of both mechanical and electrical systems. These advanced communication systems offer enhanced levels of control and reliability, but they can be affected by environmental wear and tear, leading to faults or failures. Due to their integration across various vehicle systems, issues in one area can impact multiple areas of the vehicle. Accurate diagnosis of networked architecture is essential, and oscilloscopes can contribute significantly to this. Serial decoding of data, although not available on all oscilloscopes, can provide valuable insights into the function and operation of networked systems. This section will focus on CAN BUS, one of several types of in-vehicle networks used by manufacturers.

Controller Area Networks (CAN BUS systems), enable the efficient multiplexing of data crucial for operational coherence across mechanical and electrical components.
Serial decoding in CAN BUS systems allows you to analyse and interpret the data transmitted between different **nodes** in a vehicle's network. This process involves converting serial data from the CAN BUS into a readable format that can be examined for errors and irregularities. Here's a look at how serial decoding works and how you can apply these techniques for effective diagnostics.

Advanced Techniques

Network - a communication system within a vehicle that enables data exchange between electronic control units (ECUs), sensors, and actuators.

Multiplexing - a method of transmitting multiple signals over a shared communication channel, allowing electronic control units (ECUs) and sensors to exchange data efficiently.

Controller Area Networks CAN BUS - a robust vehicle communication network that allows electronic control units (ECUs) to exchange data without a central computer.

Serial decoding - the process of interpreting serial communication signals from vehicle data networks, such as CAN, LIN, or FlexRay. It helps diagnose faults, analyse sensor data, and verify ECU communication by extracting meaningful information from raw data streams.

Node - from the Latin word 'nodus' meaning knot; individual electronic control units (ECUs), sensors, or actuators within a vehicle network that communicate via protocols like CAN, LIN, or FlexRay.

Introduction to CAN BUS data

The data sent over CAN BUS is **digital**, structured as a series of bits (0s and 1s). These bits are grouped into messages containing information like sensor readings, switch positions, commands, or system status.
To properly diagnose and understand what's being sent on the CAN BUS—especially when using oscilloscopes with CAN decoders, you must understand how this binary data is represented in different number systems.

When viewing or decoding CAN BUS data, you'll most often see values displayed in one of these three formats:

1. **Decimal** (Base 10)
    - This is the standard number system humans use daily.
    - It uses ten digits: 0 through 9.
    - Example: 255 in decimal simply means 255 units.

Use in automotive: Sensor values, voltages, RPMs, temperatures, etc., are typically displayed in decimal on scan tools.

2. **Binary** (Base 2)
    - The language of computers and ECUs.
    - Uses only two digits: 0 and 1.
    - Each digit is called a bit.
    - Example: The binary number 11111111 is equal to 255 in decimal.

Use in automotive:
    - CAN messages are transmitted in binary at the electrical level.
    - Each pulse seen on the oscilloscope corresponds to a 0 or 1.
    - Used internally by CAN decoders before conversion to more readable formats.

3. **Hexadecimal** (Base 16)
    - A shorthand for binary data, often used in diagnostics and engineering tools.
    - Uses sixteen digits: 0–9 and A–F (where A=10, B=11, ..., F=15).
    - One hexadecimal digit represents four binary bits.
    - Example: FF in hex = 11111111 in binary = 255 in decimal.

Use in automotive:
    - CAN IDs and raw data bytes are commonly displayed in hex on oscilloscopes, scan tools, or data loggers.
    - Easier to read than long binary strings but still close to the actual transmitted data.

Advanced Techniques

While oscilloscope CAN decoders often show raw hex values, always remember:

- The actual communication is in binary.
- Hex is just a compact, readable form of that binary.
- You may need to convert hex to decimal to understand what a data byte means (e.g., converting 0x1E to 30 km/h).

Quick conversion table		
Decimal	Binary	Hexadecimal
0	00000000	00
1	00000001	01
10	00001010	0A
15	00001111	0F
16	00010000	10
255	11111111	FF

Digital - data or signals represented as discrete values, typically 0s and 1s in binary form.

Binary - a numbering system that uses only two digits: 0 and 1.

Decimal - a number system based on powers of ten, commonly used in mathematics and everyday calculations.

Hexadecimal - a base-16 numeral system that uses 16 symbols: the digits 0–9 and the letters A–F to represent values 10–15.

Setting up serial decoding

To begin with serial decoding, the oscilloscope software must be capable of handling the specific protocol used in the vehicle's CAN BUS system.

Diagnostic trouble codes (DTC) will often provide a potential location for a network fault which can help with the positioning of the oscilloscope probes. Look for codes that point to specific problems, as lots of non-communication codes can be generated as a result of the fault and can be misleading.
If possible, connect to the CAN network near to the location of the suspected fault, however, if this is not possible, pins 6 and 14 of the data link connector (DLC) will often give access to CAN High and CAN Low. Be aware that some CAN communication may be blocked at the DLC due to a gateway.

Advanced Techniques

Follow these steps to set up serial decoding.

Setup and use

Step 1
- Connecting Probes: Attach the oscilloscope probes to the CAN High and CAN Low lines. These connections allow the oscilloscope to capture the differential signals transmitted across the network.

Step 2
- Start Communication: Switch on the vehicle/network and adjust time base and amplitude to acquire a stable capture. Pre-sets or guided tests can often help with this step and then be fine-tuned for accurate signal acquisition.

Step 3
- Protocol Selection: Navigate to the serial decoding menu and select the CAN protocol on the oscilloscope.
- Create new serial decoder for both CAN High and Can Low on two separate channels. Most oscilloscopes will have the parameters pre-set for the protocols used.

Step 4
- Ensure that the oscilloscope is configured to the correct bitrate and sample rate, aligned with the vehicle's CAN BUS specifications. Most oscilloscopes will have this information pre-populated but can be adjusted later if necessary.

Step 5
- Triggering: If possible, set triggering parameters to capture specific data frames or error conditions. This step ensures that the oscilloscope records relevant data without being overwhelmed by continuous traffic.

As the serial decoding available from an oscilloscope is reliant on a clean waveform image, it can sometimes misinterpret information due to signal interference.

If a serial decoder is set for both CAN High and CAN Low, the data should match. Any discrepancy can indicate a corrupt signal.

As a CAN BUS network relies on potential difference between High and Low rather than a simple on/off signal, it is relatively immune to noise. Therefore, it can sometimes be beneficial to decode from a waveform created using a maths channel as this can often cancel out network noise. This will involve implementing a maths channel that subtracts CAN Low from Can High.

Example:
If CAN High is channel A and CAN Low is Channel B the equation will be A – B.
Once complete, create a single serial decoder for this new channel.

Advanced Techniques

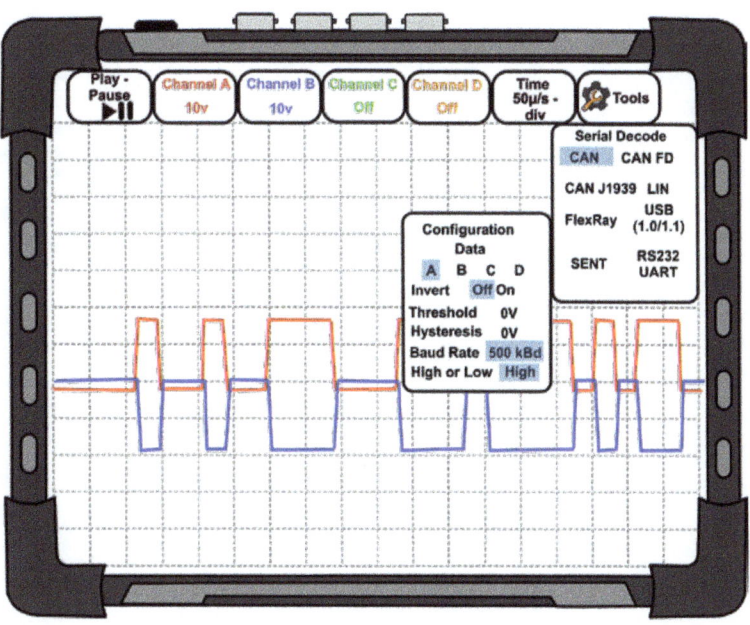

Figure 6.6 Serial decoding CAN BUS

Deciphering the data

Once the oscilloscope is set up, it begins to capture and decode the serial data from the CAN BUS. Deciphering this data involves understanding the structure of CAN messages:

- CAN Frames: Each CAN frame consists of several fields including the Identifier, Control Field, Data Field, CRC Field, ACK Field, and End of Frame. Recognising these fields is crucial for interpreting the message content. (*See Chapter 5*).
- Identifiers: The Identifier field specifies the priority and purpose of the message. It can reveal which node transmitted the data and its intended recipient.
- Data Field: The Data Field contains the actual information being communicated, such as sensor readings or control commands.
- Error Detection: The CRC Field and ACK Field are used for error detection and acknowledgment, indicating whether the message was received correctly.

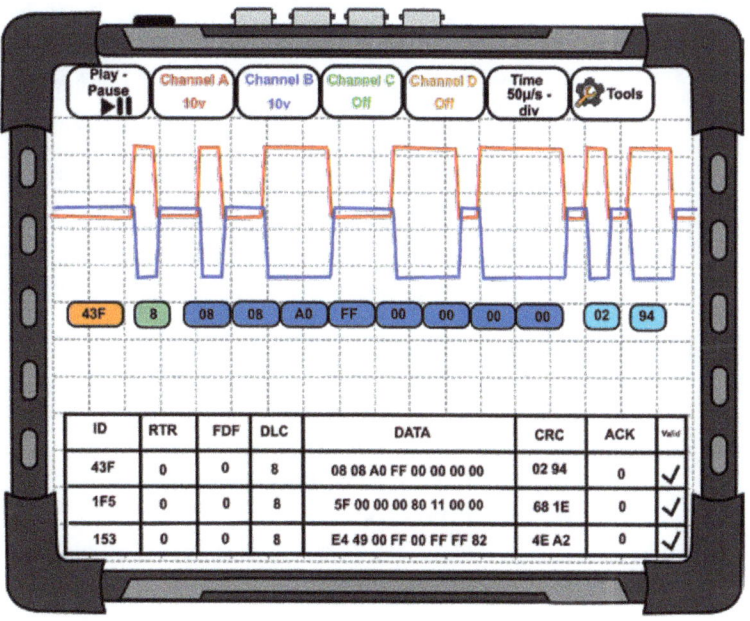

Figure 6.7 Decoded data example

Advanced Techniques

Using a Maths Channel to Calculate Potential Difference Between CAN H and CAN L
Oscilloscope Setup:
- Channel A → Connect to CAN High (CAN H)
- Channel B → Connect to CAN Low (CAN L)
  (Both measured with respect to chassis ground)

In a properly functioning CAN system:
CAN H and CAN L are differential signals — they mirror each other.
Typical values during transmission:
- CAN H rises to ~3.5V
- CAN L drops to ~1.5V
- Potential difference ≈ 2V

By calculating the instantaneous difference between CAN H and CAN L using a maths channel, you get a clean, direct representation of the true signal — noise common to both lines (common-mode noise) are naturally cancelled out.

Maths Channel Formula:
To calculate the potential difference:
A - B
Where:
- A = CAN H
- B = CAN L

This equation gives you the differential voltage, i.e., the signal the CAN transceiver actually sees.

Why This Is Useful for Diagnostics:
- Noise Immunity Check:
  Noise often appears on both CAN H and CAN L simultaneously (common-mode). By subtracting the two, the maths channel shows only differential changes — which are the actual data bits. Any common-mode spikes disappear, helping you dismiss harmless EMI.
- Quick Fault Confirmation:
  If your maths channel shows inconsistent or low-amplitude signals (far from the expected ~2V swing), you may have:
     A wiring issue
     A termination resistor problem
     A faulty module pulling one line down
- Visual Simplicity:
  Instead of visually comparing two complex waveforms, you get a single, stable waveform showing the true data transmission.

Because serial decoding using an oscilloscope works by analysing what it can see on the screen, if the data packet's frame is split (i.e., finishes on the next screen), it will see this as an error, recording this as a CRC fault. If you are receiving lots of CRC errors, try reducing the time base to see if it improves.

Practical application: diagnosing faults

You can use serial decoding to pinpoint faults within the CAN BUS physical layer; this includes both control units and wiring. CAN BUS problems mainly present as live issues and can often be diagnosed by isolating (unplugging) suspected faulty components while observing the effect on the network via the oscilloscope.

Advanced Techniques

> With serial decoding, you don't necessarily need to understand the data that is being translated. However, numerical information often makes it easier to identify errors; look for reoccurring CRC failures.
> Although CAN identifiers are proprietary and will differ from vehicle to vehicle, an internet search can often help isolate the component or area which may be causing an issue.

Case Studies

This section will provide examples of real-world scenarios where oscilloscopes play a crucial role in diagnosing complex automotive issues. **Tables 6.6 to 6.9** will cover four short case studies: Gasoline Direct Injection (GDI), camshaft/crankshaft correlation, running in-cylinder pressure transducer testing, and a CAN BUS issue.

Table 6.6 Case study

System	Case Study 1: Gasoline Direct Injection (GDI)	
Background	Gasoline Direct Injection (GDI) engines have become increasingly prevalent due to their efficiency and performance advantages. However, diagnosing faults in GDI systems can be complex due to the high-pressure fuel delivery and sophisticated control mechanisms.	Before
Diagnostic procedure	An oscilloscope is used to capture the injector waveform. The key focus is on the injector pulse, which should exhibit a distinct pattern. Deviations can indicate issues such as a damaged injector, faulty wiring, or control unit failure.	
Before and after waveforms	• Before: The waveform shows irregular injector pulses, indicating inconsistent fuel delivery. • After: Post-repair, the waveform displays regular, stable pulses, confirming the injector's proper function.	After
Outcome	The oscilloscope's readings revealed a malfunctioning injector. Replacement of the faulty injector restored normal operation, as evidenced by the corrected waveform.	

Advanced Techniques

Table 6.7 Case study

System	Case Study 2: Camshaft/Crankshaft Correlation	
Background	Camshaft and crankshaft correlation is vital for engine timing and performance. Misalignment can lead to poor engine performance, increased emissions, and potential engine damage.	Before
Diagnostic procedure	Using an oscilloscope, signals from both the camshaft and crankshaft position sensors are captured. The goal is to ensure that the signals are synchronised and within the manufacturer's specifications.	
Before and after waveforms	• Before: The waveform shows a misalignment between the camshaft and crankshaft signals, indicating timing issues. • After: Post-adjustment, the waveform displays synchronised signals confirming proper correlation.	After
Outcome	The diagnosis indicated a timing belt issue. Replacing and adjusting the timing belt realigned the camshaft and crankshaft, resulting in improved engine performance and corrected waveform synchronisation.	

Table 6.8 Case study

System	Case Study 3: Running In-Cylinder Pressure Transducer Testing	
Background	In-cylinder pressure transducers provide real-time data on the operation of the four-stroke cycle process, vital for diagnosing problems such as compression issues, misfires and performance.	Before
Diagnostic procedure	With the transducer connected to the engine in place of a spark plug, the oscilloscope captures the pressure waveform during engine operation. Key aspects include peak pressure, pressure rise rate, and pressure stability, correct valve operation and timing.	

Advanced Techniques

Table 6.8 Case study

Before and after waveforms	• Before: The waveform reveals fluctuating pressures and irregular peaks/troughs, indicative of faulty valve operation or exhaust back-pressure. • After: Post-repair, the waveform shows stable pressure readings and consistent peak and trough values.
Outcome	The oscilloscope identified a partially blocked catalytic converter as the source of the reduced engine performance. Replacing the catalytic converter restored normal pressure levels, as confirmed by the stabilised waveform.

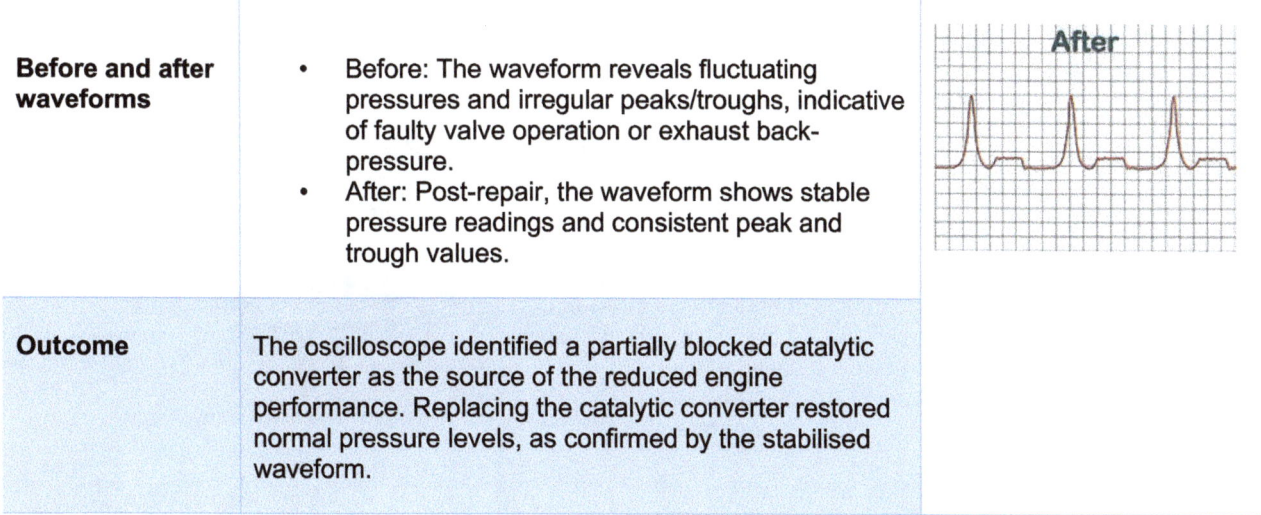

Table 6.9 Case study

System	Case Study 4: CAN BUS Issue
Background	CAN BUS networks are integral to modern vehicle electronics, facilitating communication between various control units. Faults can disrupt this communication, leading to various electronic malfunctions.
Diagnostic procedure	An oscilloscope captures CAN BUS frames, focusing on fields such as the Identifier, Control Field, Data Field, CRC Field, ACK Field, and End of Frame. The goal is to identify any inconsistencies or errors in these fields.
Before and after waveforms	• Before: The waveform displays frequent CRC errors and irregular data fields, indicating communication faults. • After: Post-diagnosis, the waveform shows corrected data fields and no CRC errors, confirming restored communication.
Outcome	The diagnosis revealed a faulty control unit causing the CAN BUS errors. Isolating and replacing the control unit resolved the communication faults, as validated by the error-free waveform.

Advanced Techniques

Educational Insights

Best practices for learning oscilloscope diagnostics

Using an oscilloscope for automotive diagnostics requires understanding waveform analysis to identify issues within vehicle systems. Effective interpretation techniques help you diagnose potential problems accurately. Here are some best practices to elevate your proficiency in oscilloscope diagnostics:

- Familiarise Yourself with Oscilloscope Functions: Start by thoroughly understanding the basic functions of your oscilloscope. Learn how to set up and adjust parameters such as time base, voltage scale, trigger settings, and sample rate. Knowing these functions will allow you to obtain accurate readings.
- Practice with Known Good Signals: Begin your training by analysing waveforms from a properly functioning vehicle. This establishes a baseline understanding of normal waveforms and helps you recognise deviations that indicate faults.
- Utilise Manufacturer Resources: Many vehicle manufacturers provide detailed waveform diagrams and diagnostic procedures specific to their models. Use these resources to gain insights into expected waveforms and typical issues.
- Participate in Training Programs: Enrol in courses or workshops focusing on automotive oscilloscopes. These programs often provide hands-on experience and expert guidance, accelerating your learning curve.

Figure 6.8 Participate in training programmes

Common missteps and troubleshooting tips

While learning to use an oscilloscope, it's essential to be aware of common pitfalls that can hinder your analysis. Avoiding these mistakes ensures accurate diagnostics and effective repairs:

- Incorrect Setup: Improperly configuring the oscilloscope can produce misleading data. Always double-check your settings before capturing waveforms. Verify the time base, voltage scale, and trigger settings align with the signal you're analysing.
- Oversampling or Under sampling: Sampling rate discrepancies can distort waveform representation. Ensure your oscilloscope's sample rate is appropriate for the signal frequency to capture accurate waveforms.
- Ignoring Noise and Interference: External electrical noise can corrupt waveform data. Use filtering options available on your oscilloscope to reduce noise and focus on the relevant signal.
- Misinterpreting Data: Incorrectly analysing waveforms can lead to misdiagnosis. Cross-reference waveforms with manufacturer-provided diagrams and consult experienced colleagues if uncertain.
- Neglecting Calibration: Regularly calibrate your oscilloscope and probes to maintain their accuracy. An uncalibrated device can give erroneous readings, compromising your diagnostics.

Advanced Techniques

Developing diagnostic strategy

An effective diagnostic routine should always begin with a logical assessment of the symptoms and then uses reasoning to reduce the possible number of options, before following a systematic approach to finding and fixing the root cause.

By integrating waveform analysis into your diagnosis, you enhance your ability to provide optimal vehicle maintenance.

Setup and use

Step 1
- Define the Problem: Begin by clearly identifying the symptoms and suspected issues. Gather information from the vehicle's owner and any diagnostic codes present.

Step 2
- Capture Reference Waveforms: Obtain waveforms from similar systems or known good parts of the vehicle. These reference waveforms serve as benchmarks for comparison.

Step 3
- Analyse Variations: Compare the captured waveforms to the reference waveforms. Note any discrepancies such as irregular shapes, unexpected spikes, or missing data fields.

Step 4
- Pinpoint Fault Location: Use the identified variations to trace back the source of the issue. Determine whether the fault is within the sensor, actuator, wiring, or control unit.

Step 5
- Validate Repairs: After performing the necessary repairs, capture new waveforms to ensure the fault has been resolved. Confirm the waveform matches the expected reference without abnormalities.

Step 6
- Document and record results: Annotated waveforms serve as excellent documentation for repairs and can be referenced in the future if similar issues arise.

Kidlin's Law states:
'If you want to understand something, write it down.'

In automotive diagnostics, this principle reinforces the importance of structured documentation. Writing down symptoms, test conditions, and oscilloscope findings forces a deeper examination of the problem, often leading to new insights or solutions that might otherwise be missed.

Mastering oscilloscope diagnostics involves continuous practice and learning. Stay updated with emerging technologies and advancements in oscilloscope functionality to refine your skills.

Advanced Techniques

Future Technologies

AI integration in oscilloscope diagnostics

As the automotive industry evolves, so does the technology used to diagnose and maintain vehicles. One of the most promising advancements is the integration of artificial intelligence (AI) in oscilloscope diagnostics. AI has the potential to revolutionise the way technicians interpret waveforms, making the diagnostic process faster and more accurate.

Traditionally, waveform analysis requires a technician to manually compare captured waveforms to reference waveforms, noting any discrepancies and pinpointing the source of faults. This process, while effective, can be time-consuming and dependent on the skill level of the mechanic. AI can streamline this process by using machine learning algorithms to automatically interpret waveforms and identify anomalies with greater precision.

AI can be trained on vast datasets of waveform patterns, learning to recognise the signatures of various faults. This enables it to provide real-time analysis and recommendations, reducing the margin for error and enhancing the efficiency of diagnostics. For technicians, this means spending less time on manual interpretation and more time on performing repairs, ultimately improving the quality of service provided to vehicle owners.

Moreover, AI integration might provide predictive maintenance by identifying patterns that indicate a potential future failure. By detecting these early warning signs, technicians can proactively address issues before they become critical, thereby extending the lifespan of vehicle components and preventing breakdowns.

However, we must not rely solely on artificial intelligence, and it should perform a complementary service to the skills of the technician.

Next-generation communication protocols

The advent of next-generation communication protocols, such as Ethernet, is transforming automotive diagnostics. Modern vehicles are increasingly equipped with complex electronic systems that require robust and efficient communication networks. Ethernet, known for its high-speed data transfer and reliability, is emerging as a key player in automotive networks.

Ethernet-based systems offer several advantages over traditional protocols. They provide faster data rates, which are essential for handling the large volumes of information generated by advanced sensors and control units. This speed is crucial for real-time diagnostics, enabling you to quickly access and analyse data from multiple systems within the vehicle.

Also, Ethernet supports greater scalability and flexibility. As automotive technology continues to advance, the ability to integrate new systems seamlessly is vital. Ethernet's standardised architecture ensures compatibility with a wide range of devices, making it easier to upgrade and expand the diagnostic capabilities of a vehicle.

Preparing for the integration of Ethernet in automotive networks involves understanding its operational principles and implementing the necessary infrastructure. You must familiarise yourself with Ethernet protocols, network configurations, and troubleshooting techniques to effectively use this technology in diagnostics.

The transition to Ethernet also calls for updated diagnostic tools. Oscilloscopes and other diagnostic equipment must be compatible with Ethernet-based systems to accurately capture and interpret data. Investing in these advanced tools will position you to stay ahead of technological trends and provide superior diagnostic services.

By embracing AI and next-generation communication protocols, you can significantly enhance your diagnostic capabilities. These emerging technologies offer the potential to improve accuracy, efficiency, and predictive maintenance in automotive diagnostics. As the industry continues to evolve, staying informed and adaptable will be key.

Conclusion

While it is impractical to try and cover all possible scenarios, waveforms, settings, and issues, this book seeks to provide comprehensive insights into many functions and applications of an automotive oscilloscope. The goal is to develop your confidence in the setup and operation of oscilloscopes and foster a sense of curiosity that will motivate you to explore and utilise various features within your own systematic diagnostic procedures.

Common Acronyms/Abbreviations

This section contains common acronyms and abbreviations: An acronym is a word that is formed from the first letters of a phrase or a series of words, usually to make it easier to say or remember. This list is not exhaustive but provides some acronyms used in conjunction with the design and operation of electric vehicles. Abbreviations may have different meanings or designations depending on context, and acronyms may be further adapted, reused, or reinterpreted as technology and engineering develops.

A - Amperes
A/F - Air Fuel Ratio
A/T - Automatic Transmission
AAT - Ambient Air Temperature
ABS - Antilock Brake System
AC - Alternating Current
AC - Air Conditioning
ACC - Automatic Climate Control
ACC - Air Conditioning Clutch
ACR - Air Conditioning Relay
ACR4 - Air Conditioning Refrigerant, Recovery, Recycling, Recharging
ADU - Analogue-Digital Unit
ADC - Analogue to Digital Converter
AED - Automatic Electronic Defibrillators
AEV- All Electric Vehicle
AFR - Air Fuel Ratio
AGM- Absorbed Glass Matt
Ah- Amp Hours
AM - Amplitude Modulation
APP - Accelerator Pedal Position
ATS - Air Temperature Sensor
AVO- Amps Volts Ohms
AWG - American Wire Gage
BBW - Brake by Wire
BCM - Body Control Module
BCM - Battery Control Module
BEV - Battery Electric Vehicle
BHP - Brake Horsepower
BMS - Battery Management System
BMU - Battery Management Unit
BNC - Bayonet Neill-Concelman
BOB - Breakout Box
BPP - Brake Pedal Position Switch
BTS - Battery Temperature Sensor
BUS N - Bus Negative
BUS P - Bus Positive
C - Celsius
C - Coulomb
CA - Cranking Amps
CAN - Controller Area Network
CAT - Category

CC - Catalytic Converter
CC - Climate Control
CC - Cruise Control
CCA - Cold Cranking Amps
CCS - Combined Charging System
CCV - Closed Cricut Voltage
CL - Closed Loop
CLV - Calculated Load Value
CNG - Compressed Natural Gas
CNP - Coil Near Plug
CO - Carbon Monoxide
CO2 - Carbon Dioxide
COP - Coil on Plug
COSHH - Control of Substances Hazardous to Health
CP - Control Pilot
CPR - Cardiopulmonary Resuscitation
CPU - Central Processing Unit
CRC - Cyclic Redundancy Check
CTP - Closed Throttle Position
CTS - Coolant Temperature Sensor
CV - Constant Velocity
CVT - Continuously Variable Transmission
DBW - Drive by Wire
DC - Duty Cycle
DC - Direct Current
DMM - Digital Multimeter
DLC - Data Link Connector (OBD)
DPF - Diesel Particulate Filter
DSO - Digital Storage Oscilloscope
DTC - Diagnostic Trouble Code
EBCM - Electronic Brake Control Module
EBD - Electronic Brake Force Distribution
ECM - Engine/Electronic Control Module
ECS - Emission Control System
ECT - Engine Coolant Temperature
ECU - Electronic Control Unit
EECS - Evaporative Emission Control System
EEGR - Electronic EGR (Solenoid)
EEPROM - Electronically Erasable Programmable Read Only Memory
EGO - Exhaust Gas Oxygen Sensor
EGR - Exhaust Gas Recirculation
EGRT - Exhaust Gas Recirculation Temperature
EMF - Electromotive Force (voltage)
EMI - Electromagnetic Interference
EML - Engine Management Light
EOBD - European Onboard Diagnostics
EPA - Environmental Protection Act
EPB - Electronic Parking Brake
EPB - Equipotential bonding
EPROM - Erasable Programmable Read Only Memory
EPS - Electronic Power Assisted Steering

Common Acronyms/Abbreviations

ESP - Electronic Stability Programme
ESS - Engine Start-Stop
EV - Electric Vehicle
EVAP - Evaporative Emissions System
EVAP CP - Evaporative Canister Purge
EVSE - Electric Vehicle Supply Equipment
F - Farad
F – Fahrenheit
FFT - Fast Fourier Transform
FM - Frequency Modulation
FOT - Fixed Orifice Tube
FSD - Full Scale Deflection
FWD - Front Wheel Drive
GND - Electrical Ground Connection
H - Hydrogen
H - Henry
HASAWA - Health and Safety at Work Act
HC - Hydrocarbons
HCA - Hot Cranking Amps
HEGO - Heated Exhaust Gas Oxygen Sensor
HEV - Hybrid Electric Vehicle
HFC - Hydrogen Fuel Cell
HICE - Hydrogen Internal Combustion Engine
HO2S - Heated Oxygen Sensor
hp - Horsepower
HSE - Health and Safety Executive
HT - High Tension
HV - High Voltage
HVAC - Heating Ventilation and Air Conditioning
HVIL - High Voltage Interlock Loop
Hz – Hertz
I - Intensité du courant
I/O - Input / Output
IA - Intake Air
IAT - Intake Air Temperature
IC - Integrated Circuit
ICCB - In Cable Charging Box
ICE - In Car Entertainment
ICE - Internal Combustion Engine
IGBT - Insulated Gate Bipolar Transistor
IGN - Ignition
IHKA - Climate Control (network acronym)
IMU - Inertial Measurement Unit
ISO - International Standard of Organisation
ISS - Input Speed Sensor
J - Joule
KAM - Keep Alive Memory
Kg/cm2 - Kilograms/Cubic Centimetres
kHz - Kilohertz
km - Kilometres
Kombi - Instrument Cluster (network acronym)
KPA - Kilopascal
kW - Kilowatt
KWP - Keyword Protocol

l - Litres
LA - Lead Acid
LCD - Liquid Crystal Display
LED - Light Emitting Diode
LEV - Low Emission Vehicle
LFP - Lithium Iron Phosphate
LHD - Left Hand Drive
Li-ion - Lithium ion
LIN - Local Interconnect Network
LMO - Lithium Manganese Oxide
LOS - Limited Operating Strategy
LPG - Liquefied Petroleum Gas
LTO - Lithium Titanate
LWB - Long Wheelbase
LWR - Vertical Headlight Control (network acronym)
M/T - Manual Transmission
MAC - Mobile Air Conditioning
MAF - Mass Air Flow
MAP - Manifold Absolute Pressure
MCM - Motor Control Module
MEF - Methane Equivalency Factor
MF - Maintenance Free
MIL - Malfunction Indicator Lamp
MOSFET - Metal-Oxide-Semiconductor Field-Effect Transistor
MPG - Miles per Gallon
MPH - Miles per Hour
MRE - Magnetic Resistive Element
MRS - Multiple Restraint System (network acronym)
mS or ms - Millisecond
MSD - Maintenance Service Disconnect/Manual Service Disconnect
mV or mv - Millivolt
N - Newton
N - Nitrogen
NCA - Nickel Cobalt Aluminium
NCAPS - Non-Contact Angular Position Sensor
NCM - Nickel Cobalt Manganese
NCRPS - Non-Contact Rotary Position Sensor
NGV - Natural Gas Vehicles
NIB - Neodymium Iron Boron
Ni-MH - Nickel Metal Hydride
Nm - Newton Meters
NOx - Oxides of Nitrogen
NPN - Negative Positive Negative
NTC - Negative Temperature Coefficient
NVH - Noise Vibration and Harshness
O2 - Oxygen
OBC - Onboard Charger/Offboard Charger
OBD I - On Board Diagnostics Version I
OBD II - On Board Diagnostics Version II
OCV - Open Circuit Voltage
OD - Outside Diameter
ODP - Ozone Depletion Potential

Common Acronyms/Abbreviations

OE - Original Equipment
OEM - Original Equipment Manufacturer
OFN - Oxygen Free Nitrogen
OL - Off Limits
OL - Open Loop
OS - Oxygen Sensor
OSS - Output Speed Sensor
P/N - Part Number
PAG - Polyalkylene Glycol
PATS - Passive Anti-Theft System
PCB - Printed Circuit Board
PCM - Powertrain Control Module
Pd - Potential Difference (volts)
PE - Protected Earth
PEF - Propane Equivalency Factor
PEM - Proton Exchange Membrane
PEV - Pure Electric Vehicles
PH - Potential Hydrogen
PHEV - Plug-in Hybrid Electric Vehicle
PID - Parameter Identification Location
PKE - Passive Keyless Entry
PLC - Powerline Communication
PNP - Positive Negative Positive
POE - Polyolester Oil
POF - Plastic Optical Fibre
POT - Potentiometer
PP - Proximity Pilot
PPE - Personal Protective Equipment
PPM - Parts Per Million
PPS - Accelerator Pedal Position Sensor
PROM - Programmable Read-Only Memory
PSI - Pounds per Square Inch
PTC - Positive Temperature Coefficient
PTM - Pulse Train Module
PUWER - Provision and Use of Work Equipment Regulations
PWM - Pulse Width Modulation
RAM - Random Access Memory
RBS - Regenerative Braking system
RCD - Residual Current Device
RCM - Reserve Capacity Minutes
RCM - Restraint Control Module
RDS - Radio Data System
RDW - Tyre Pressure Monitoring (network acronym)
RE EV - Range Extended Electric Vehicles
REF - Reference
RESS - Rechargeable Energy Storage System
RFI - Radio Frequency Interference
RHD - Right Hand Drive
RIDDOR - Reporting of Injuries Diseases and Dangerous Occurrence Regulations
RKE - Remote Keyless Entry
RMS - Recovery Management Station
RMS - Root Mean Square

ROM - Read Only Memory
RON - Research Octane Number
RTV - Room Temperature Vulcanizing
RWD - Rear Wheel Drive
SAE - Society of Automotive Engineers (Viscosity Grade)
SEI - Solid Electrolyte Interphase/Interface
SIPS - Side Impact Protections System
SMR - System Main Relay
SoC - State of Charge
SoH - State of Health
SPS - Samples Per Second
SRI - Service Reminder Indicator
SRS - Supplementary Restraint System (air bag)
SRT - System Readiness Test
SWB - Short Wheelbase
SWL - Safe Working Load
SZM - Central Switch Module (network acronym)
TACH - Tachometer
TCM - Transmission Control Module
TCS - Traction Control System
TP - Throttle Position
TPM - Tyre Pressure Monitor
TPP - Throttle Position Potentiometer
TPS - Throttle Position Sensor
TSB - Technical Service Bulletin
TXV - Thermal Expansion Valve
UART - Universal Asynchronous Receiver-Transmitter
UJ - Universal Joint
ULEV - Ultra Low Emission Vehicle
USB - Universal Serial Bus
UV - Ultraviolet
V - Volts
V2G - Vehicle to Grid
VAC - Vacuum
VDU - Visual Display Unit
VDE - Verband der Elektrotechnik
VIN - Vehicle Identification Number
VPE - Vehicle Protection Equipment
VR - Variable Reluctance
VSS - Vehicle Speed Sensor
W/B - Wheelbase
Wh - Watt Hours
WPT - Wireless Power Transfer
WSS - Wheel Speed Sensor
WVO - Waste Vegetable Oil
ZEV - Zero Emission Vehicle

Appendix
DIN Terminal Numbers

Ignition system

1	coil, distributor, low voltage
1a, 1b	distributor with two separate circuits
2	breaker points magneto ignition
4	coil, distributor, high voltage
4a, 4b	distributor with two separate circuits, high voltage
7	terminal on ballast resistor, to distributor
15	battery+ from ignition switch
15a	from ballast resistor to coil and starter motor

Preheat (Diesel engines)

15	preheat in
17	start
19	preheat (glow)

Starter

50	starter control

Battery

15	battery+ through ignition switch
30	from battery+ direct
30a	from 2nd battery and 12/24 V relay
31	return to battery- or direct to ground
31a	return to battery- 12/24 V relay
31b	return to battery- or ground through switch
31c	return to battery- 12/24 V relay

Electric motors

32	return
33	main terminal (swap of 32 and 33 is possible)
33a	limit
33b	field
33f	2. slow rpm
33g	3. slow rpm
33h	4. slow rpm
33L	rotation left
33R	rotation right

Turn indicators

49	flasher unit in
49a	flasher unit out, indicator switch in
49b	out 2. flasher circuit
49c	out 3. flasher circuit
C	1st flasher indicator light
C2	2nd flasher indicator light
C3	3rd flasher indicator light
L	indicator lights left
R	indicator lights right
L54	lights out, left
R54	lights out, right

AC generator (alternator)

51	DC at rectifiers
51e	as 51, with choke coil
59	AC out, rectifier in, light switch
59a	charge, rotor out
64	generator control light

Generator, voltage regulator

61	charge indicator (charge control light)
B+	battery +
B-	battery -
D+	dynamo +
D-	dynamo -
DF	dynamo field
DF1	dynamo field 1
DF2	dynamo field 2
U, V, W	AC three phase terminals

Appendix
DIN Terminal Numbers

Lights

54	brake lights
55	fog light
56	spot light
56a	headlamp high beam and indicator light
56b	low beam
56d	signal flash
57	parking lights
57a	parking lights
57L	parking lights left
57R	parking lights right
58	licence plate lights, instrument panel
58d	panel light dimmer

Window wiper/washer

53	wiper motor + in
53a	limit stop+
53b	limit stop field
53c	washer pump
53e	stop field
53i	wiper motor with permanent magnet, third brush for high speed

Acustic warning

71	beeper in
71a	beeper out, low
71b	beeper out, high
72	hazard lights switch
85c	hazard sound on

Switches

81	opener
81a	1 out
81b	2 out
82	lock in
82a	1st out
82b	2nd out
82z	1st in
82y	2nd in
83	multi position switch, in
83a	out position 1
83b	out position 2

Relay

85	relay coil -
86	relay coil +

Relay contacts

87	common contact
87a	normally closed contact
87b	normally open contact
88	common contact 2
88a	normally closed contact 2
88b	normally open contact 2

Additional

52	signal from trailer
54g	magnetic valves for trailer brakes
75	radio, cigarette lighter
77	door valves control

Index

A

Accessories · 40
Acquiring waveforms · 66
acronyms · 211
Actions · 186
Actuators · 75
AI · 210
Algorithms · 96
Aliasing · 35, 101
Alternating · 19
alternator · 80, 162
Ambient temperature · 81
Amplifier · 95
Amplitude · 27, 35, 38, 39, 47, 95, 129, 182, 211
Amps · 11
Analogue · 32, 95, 211
Analogue oscilloscopes · 32
analogue-to-digital converter · 94
Annotating · 96, 114, 121
Anomalies · 96
Architecture · 93, 94, 95
Atoms and molecules · 8
Attenuator · 44, 59, 95
Auto trigger · 97

B

Back EMF · 59
Back-probe · 34, 43
Bandwidth · 37
Basic Configurations · 53
battery · 18
Battery power · 52
Bayonet · 43, 211
Benchtop oscilloscopes · 32
Best practices · 180
Binary · 200, 201
Bits · 38, 101
BNC · 41, 43, 44, 211
Breakout · 34
Brushes · 163, 164
Buffers · 57

C

Calibration · 44, 59, 62, 85, 108, 208
camshaft · 71
CAN BUS · 59, 81, 82, 87, 114, 133, 134, 135, 136, 139, 140, 142, 144, 171, 187, 193, 198, 199, 200, 201, 202, 203, 204, 205, 207
Capacitance · 13, 15
Case Studies · 205
Cathode-ray-tube · 40
CATIII · 62, 91, 104
Celsius · 14
Channel setup · 58
Channels · 33
Charge · 13, 15, 213
charging system · 162
Chemical Electricity · 18
Circuit · 7, 10, 16, 18, 212, 213
circuit loop · 42
Closed-circuit voltage · 15
Coil-on-plug · 43
Common-mode voltage · 103, 104
Conductor · 10
Contactors · 145
Continuity · 10, 18
Control panel · 38, 44
Controller area network · 30, 81, 133, 134, 211
controls · 44
Coulombs · 13
crankshaft · 71
Crocodile clips · 41
Current · 7, 8, 11, 19, 20, 21, 22, 23, 24, 75, 192, 194, 195, 199, 211, 213
Cursors · 47, 48, 62, 63, 64
Custom probe · 108, 172
Custom Signal Capture · 171
Cycles · 40, 48

D

data link connector · 136
Data packet · 134
Data transmission · 136
Decimal · 200, 201
Deep measure · 183
diagnostic scan tools · 111
diagnostic strategy · 209
Differential amplifier · 104
Differential measurement · 149
Differential probe · 35, 41, 103, 104
Differential voltage · 35, 69
Digital · 32, 36, 85, 95, 178, 201, 211
Digital oscilloscopes · 32
Digitised · 95
Diode · 163, 212
Direct · 19
Display · 38

Index

Display Interface · 46
Display resolution · 37
dropouts · 120
Dual screen view · 54
Duty cycle · 25, 126, 127, 128, 129, 153
Dwell period · 162
Dynamic Testing · 175

E

earth · 16, 24, 26, 34, 149
Edge matching · 59
Educational Insights · 208
EGR valves · 76
Electric motor control · 132
Electric shock · 20
electric vehicle supply equipment · 149, 151
Electric vehicles · 88
Electrical noise · 52
Electrical Units and Terminology · 11
Electromagnetism · 18
Electromotive Force · 11, 15, 23, 211
Electron · 9, 10
Electronic and electrical safety procedures · 5
Error Analysis · 199
EV · 144
EVSE · 149, 150, 151, 152, 153, 212
Exciter · 148
exhaust · 83, 166
Extraneous signals · 69
Eye protection · 90

F

Fahrenheit · 14
Farads · 13
Fast Fourier Transform (FFT) · 94, 96
Filtering · 85, 109, 196
Filters · 96, 109
First look · 68, 69
floating ground · 34, 149
Floating signals · 104
footwear · 91
Frequency · 12, 15, 27, 34, 38, 39, 40, 47, 85, 128, 155, 175, 181, 197, 212, 213
frequency modulation · 155
Fuel Injector · 130
Fuel injectors · 75
Future Technologies · 210

G

glitches · 120

Granularity · 38
Graticule · 48
grid · 46
Ground · 16, 43, 104, 212
Grounding · 29
Guided tests · 118

H

Hall effect · 42, 44
Handheld oscilloscopes · 31
Handling and care · 44
Henry · 13
Hertz · 12
Hexadecimal · 200, 201
High Voltage · 22
High-tension · 43
How to use this book · 3
Hybrid · 144

I

IGBT · 129, 130, 155, 212
Ignition system · 131, 160
Ignition waveforms · 70
Impedance · 58, 59
In-cylinder compression · 83
Inductance · 13, 15
Inductive · 42, 44
Inductive amps measurement clamps · 42
Inductive secondary ignition probes · 42
Inertia · 81
Information sources · 5
Initial Setup · 50
Injector · 157
Inlet · 166
Insulated tools · 107
Insulator · 10
Intake · 83
Interference · 35, 43, 85, 208, 211, 213
intermittent faults · 176
internet forums · 117
Interpreting waveforms · 68
In-vehicle networks · 81
Isolation · 105

J

Joule · 14, 15, 212
Joules · 14

Index

K

Keyless entry detector · 42
Known good · 69, 115

L

lab scope · 32
Laptop-based DSOs · 32
leads · 38, 40, 41, 52, 68, 87, 103, 118, 177
Lighting · 132
low voltage · 22, 46, 60, 89, 106, 214

M

MAF · 71
Mains power · 52
Maintenance · 85
Masks · 184
Maths Channels · 120, 191
Mbps · 133, 134
Measurement data · 47
measurement units · 174
Measurements · 181
Mediation · 134
Memory Depth · 102
Menus · 45
missteps · 208
Molecule · 8
MOSFET · 129, 130, 212
Motor controllers · 154
Movement of electrons · 9
multimeters · 112
Multiplex · 134, 200
Multi-screen view · 54

N

Network · 30, 133, 134, 138, 200, 212
Newton · 15, 212
Node · 134, 200
noise · 108
Noise vibration and harshness · 169
Noise, vibration, and harshness (NVH) probes · 43
Nucleus · 8
NVH · 84

O

O2 sensors · 71

Ohms · 12
Ohms law · 23
Open circuit voltage · 15
Oscillate · 15
Oscilloscope · 29
Oscilloscope maintenance · 86
Overalls · 90
Overlay view · 55
Overlays · 121
Oxygen sensors · 73

P

Parallel · 17, 18, 22, 137
Parasitic draw · 81
Passive probes · 41
Pattern Recognition trigger · 99
periodic table · 9
Personal Protective Equipment (PPE) · 6, 89
Physical layer · 133, 134
Pixels · 38
Portable oscilloscopes · 31
Potential Difference · 11, 15, 204, 213
Power · 7, 12, 15, 21, 23, 24, 33, 49, 52, 77, 88, 166, 182, 195, 199, 212, 213
Preparing for assessment · 4
Presets · 118
pressure transducer · 43, 83, 164
Probes · 27, 29, 38, 40, 41, 86, 87, 170
Protocols · 36, 95, 134, 170, 212
Proton · 10, 213
Pull-up and Pull-down Circuits · 26
Pulse Width Diagnostics · 126
Pulse width modulation · 25, 126, 127, 128
Pulse width trigger · 98

Q

Quick Set-up Guide · 29

R

Ready mode · 149
Rectifier · 81
Rectify · 164
Reference waveforms · 115, 188
Relative compression · 78
relays · 145
Repeat trigger · 98
Resistance · 15
Resolution · 35, 36, 37, 101
Resolvers · 147

Index

Ringing · 162
Ripple · 163
RMS · 19
root-mean-square · 19
Rulers · 48, 62, 63, 64
Runt signals · 96, 98
Runt trigger · 98

S

Safety Considerations · 87
Sample rate · 36
Sampling · 35, 100
Saving and exporting data · 69
Scope · 29, 174
Secondary ignition connectors · 41
Semiconductor · 10, 130, 212
Sensor diagnostics · 71
Serial · 134, 199, 200, 203, 213
serial decoding · 201
Series · 17, 18
Shielded probes · 69
Short circuit · 29, 89
Signal fidelity · 101
Signal filtering · 15, 110
Sinewave · 46, 148
Single screen view · 53
Single trigger · 98
social media · 117
Spark Plug · 70
Split-screen view · 56
starter motor · 78
static electricity · 8
Sweep · 27, 34, 38, 39, 65
Synchronisation · 59, 111, 125, 127

T

Temperature · 14, 15, 75, 164, 211, 212, 213
Throttle bodies · 77
Time and voltage settings · 60

Time base · 38, 39
Timing Measurements · 125
Topology diagrams · 134
Trace · 40
Transducer · 44, 206
Transient events · 36, 65
Transistor · 130, 155, 212
Trigger · 45, 64, 65, 66, 94, 96, 97, 98, 99, 187
Troubleshooting · 85, 208
Turbocharger · 131

U

Ultrasonic · 42, 44, 75
Ultrasonic parking sensor detector · 42
USB power · 52

V

Variable reluctance · 149
Vehicle Protective Equipment (VPE) · 6
Views · 53
Volts · 11

W

Watt's (power) law · 24
Watts · 12
Waveform Libraries · 115
Waveforms · 29, 46
What is an Oscilloscope · 28
What is Electricity · 8
workwear · 90

Z

Zooming · 40, 96

Other Books

If you enjoyed this book, please check out other titles by the author:

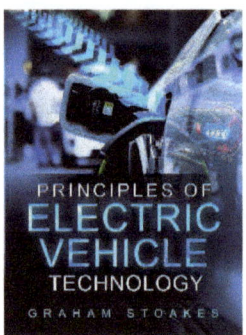

Paperback: 238 pages
Publisher: Graham Stoakes
(05 August 2024)
ISBN-10: 0992949270
ISBN-13: 978-0992949273

Principles of Electric Vehicle Technology

Electric vehicles (EVs) are a vital part of the world's infrastructure and are a fundamental component in the ongoing development that powers our future transportation needs.

Limited resources and environmental pollution mean that an alternative to traditional petrol- and diesel-powered internal combustion engines is necessary to maintain and propel a sustainable society that continues to thrive.

Regardless of fuel, it is electricity that has an elemental energy which produces no toxic pollutants at point of use.

Electricity can be generated from multiple sources, many of which can be considered environmentally friendly, making it a logical choice to charge our energy needs.

In our interconnected modern lives, electricity is indispensable, powering everything from our homes to our gadgets. Yet even though the first cars developed used electricity for their source of propulsion, internal combustion engines dominated the automotive landscape due to their seemingly inexhaustible and harmless hydrocarbon fuel.

As the world developed, it became increasingly clear that our reliance upon petrol and diesel was misguided.

We are perfectly happy with using electricity for nearly all of our daily energy needs, but many still struggle with the concept of using it for transportation due to the abundance of the internal combustion engine.

The inevitable rise in electrically powered vehicles means that we all need to learn about this new (yet old) technology if we are to integrate it into our everyday lives.

Often, the unknown poses the greatest barrier to a change in habits, leading to mental roadblocks that hinder progress.

We often know more than we give ourselves credit for, but it takes a spark of inspiration for this to be realised.

This book is designed to support knowledge relating to the technology employed in the operation and use of electric vehicles; helping technicians, engineers, first responders and vehicle operators understand how they work.

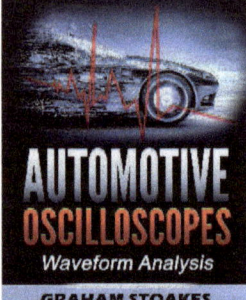

Paperback: 198 pages
Publisher: Graham Stoakes
(24 April 2017)
ISBN-10: 099294262
ISBN-13: 978-0992949266

Automotive Oscilloscopes Waveform Analysis

The rapid growth of technology used in cars has highlighted the need for a piece of diagnostic equipment that will give you X-ray vision and show you the heartbeat of a vehicles electrical and electronic system.

An OBD scan tool is vital for modern vehicle diagnostics; however, trouble codes will only take you so far. The problem can arise when the phrase 'fault code' is used in connection with diagnosis. A code will rarely point you directly to the root cause of a vehicle fault but can help you focus your diagnosis on a specific area and run functional tests.

It is the oscilloscope (or scope) that can truly test the operation and health of a system component. The important thing to remember about oscilloscopes, is that they should be easy to set up and use; otherwise, they will be passed over for a more familiar tool within your comfort zone.

Remember that nothing ever happens within your comfort zone.

There is a great deal of misconception about how difficult a scope can be to set up, and once you are used to your own equipment, if it is laid out and ready to use, it will soon become your diagnostic tool of choice.

This book has been written to help you get the most from your oscilloscope and has been designed to give straightforward and uncomplicated methods that can be used effectively for automotive diagnosis. It covers many of the most common automotive waveforms, assisting you in the analysis of the patterns produced, without restricting you to rigid equipment settings, or vehicle system design. This will give you the 'scope' to develop your systematic diagnostic routines, with the flexibility to adapt to changing requirements.

Other Books

Paperback: 338 pages
Publisher: Graham Stoakes
(1 Feb. 2015)
ISBN-10: 099294922X
ISBN-13: 978-0992949228

Level 4 Automotive Master Technician - Advanced Light Vehicle Technology
'Technology needs technicians, and the ability to harness technical diagnosis calls for a Master Technician'.
The rapid growth in technology used in the production of cars has highlighted the need for a different approach to vehicle diagnosis and repair. The integration of complex electronic control with mechanical systems shows the brilliance in the engineering capabilities of designers and manufacturers.
While this technology has improved the comfort, safety, convenience and reliability of vehicles, it has also created an issue with established methods of maintenance and repair. As many of the control systems operate beyond our natural capabilities, diagnostic tooling is required to undertake most of the fault-finding duties traditionally conducted by vehicle technicians. Also, the sophisticated nature of advanced system faults will often lead to diagnostic requirements for which there is no prescribed method.
One of the fundamental roles of a Master Technician will be the diagnosis and repair of these complex and advanced system faults, for which diagnostic approaches need to be developed that can provide logical strategies to reduce overall diagnostic time. An effective diagnostic routine should always begin with a logical assessment of symptoms and then uses reasoning to reduce the possible number of options, before following a systematic approach to finding and fixing the root cause.

Paperback: 154 pages
Publisher: Graham Stoakes
(6 July 2015)
ISBN-10: 0992949246
ISBN-13: 978-0992949242

Principles of Light Vehicle Air Conditioning
'As the number of vehicles on the world's roads rises, the demand for increased levels of comfort and convenience also grows'
While air conditioning and climate control may be seen as a luxury by some, the key benefits often outweigh the initial costs and resources required to implement these systems on newly produced vehicles; in fact most new cars come with some form of air conditioning as standard.
An environment which helps keep the driver and passengers comfortable and alert, maintaining the correct levels of ventilation and humidity, can increase concentration and the ability to devote more of their attention to the occupation of driving.
The downside of these systems is the environmental impact of the chemicals used to provide the refrigeration process.
Globally, anthropogenic, or 'man-made' emissions are believed to be the key factor in climate change and refrigerants have a larger influence than many others.
Small amounts of fluorinated gasses released to atmosphere may be causing irreparable damage to our planet, initiating ozone depletion and global warming.
Although many organisations are currently seeking alternatives to these harmful cocktails, at the present time we are restricted by the availability, cost and technology required to make viable replacements.
This means that for the time being, technicians and air conditioning professionals need to ensure that refrigerants are handled with due diligence and systems are maintained to the highest standards in order to contain and reduce emissions. Remember these chemicals only become dangerous when released to atmosphere.

Paperback: 196 pages
Publisher: Graham Stoakes
(1 July 2014)
ISBN-10: 0992949203
ISBN-13: 978-0992949204

Hybrid electric and alternative automotive propulsion
'A future without oil won't spell the end of the car but will simply drive engineering brilliance to find an alternative'.
As fuel demand and environmental pollution increases, it is important that substitutes are found for traditional methods of vehicle drive. An alternative propulsion vehicle is one that operates using something other than the established petrol or Diesel.
Whether you are a vehicle technician, automotive trainer, student or part of the emergency services, an awareness of current and emerging propulsion sources is vital to work on or around these vehicles safely.

www.grahamstoakes.com

www.ingramcontent.com/pod-product-compliance
Lightning Source LLC
Chambersburg PA
CBHW061158010526
44119CB00060B/856